Logics for New-Generation AI
Fourth International Workshop
15-16 June 2024, Hangzhou

Volume 1
Proceedings of the First International Workshop, Hangzhou, 2021
Beishui Liao, Jieting Luo and Leendert van der Torre, eds

Volume 2
Proceedings of the Second International Workshop, Zhuhai, 2022
Beishui Liao, Réka Markovich and Yì N. Wáng, eds

Volume 3
Logics for AI and Law. Joint Proceedings of the Third International Workshop on Logics for New-Generation Artificial Intelligence and the International Workshop on Logic, AI and Law, September 8-9 and 11-12, 2023, Hangzhou, Zhuhai, 2022
Bruno Bentzen, Beishui Liao, Davide Liga, Réka Markovich, Bin Wei, Minghui Xiong and Tianwen Xu, eds

Volume 4
Proceedings of the Fourth International Workshop, Hangzhou, 2024
Beishui Liao, Jun Pang and Tjitze Rienstra, eds

Logics for New-Generation AI
Fourth International Workshop
15-16 June 2024, Hangzhou

Edited by
Beishui Liao
Jun Pang
Tjitze Rienstra

ISBN 978-1-84890-459-0

College Publications, London
Scientific Director: Dov Gabbay
Managing Director: Jane Spurr

http://www.collegepublications.co.uk

Original cover design by Laraine Welch

Preface

We are delighted to introduce the proceedings of the Fourth International Workshop on Logics for New-Generation Artificial Intelligence (LNGAI2024), held in Hangzhou, China, at Zhejiang University from June 15th to 16th. This workshop falls under the umbrella of the national key project – 'Research on Logics for New Generation Artificial Intelligence' (No.20&ZD047, 2021-2025). The core aims of the LNGAI project encompass the advancement of theories and techniques in non-monotonic logics and formal argumentation, along with their applications in causal reasoning, knowledge graph reasoning, and reasoning about norms and values in the context of open, dynamic, and real-world environments. The papers compiled in this volume of proceedings highlight advancements in the interdisciplinary research of logic and artificial intelligence.

For this workshop, we have accepted a total of eight papers, each of which underwent rigorous peer-review by at least three. Furthermore, the proceedings are enriched with an invited paper from a subproject leader of LNGAI, alongside seven abstracts highlighting keynote presentations.

On one hand, the seven keynote talks highlight pivotal research trends, encompassing the integration of Large Language Models with BOID, the application of deep learning for dynamical system inference, the causality and responsibility in Multi-Agent Systems, the alignment of AI values, and the connection of prioritised default logic and structured argumentation, the explanation method in answer set programming, and the categorization for explainable AI and human-machine interaction. More specifically, Jan Broersen reviewed the BOID architecture and suggested new directions in relation to Large Language Models (LLMs). Jun Pang provided an overview of recent developments in deep learning methods for structural inference of dynamical systems, with a focus on variational auto-encoders (VAEs) based methods. Mehdi Dastani discussed the intertwined concepts of causality and responsibility in multi-agent systems, proposing a formal approach to model interactions and analyze strategic decisions. Marija Slavkovik addressed AI value alignment from the perspective of automating moral reasoning and decision-making for autonomous systems. Leendert van der Torre compared prioritised default logic (PDL) and structured argumentation based on the weakest link, proposing a new variant of PDL and analyzing it within the framework of attack relation assignments. Yisong Wang, Thomas Eiter, Yuanlin Zhang and Fangzhen Lin introduced a witness theory for answer set programs to explain the conclusions of logic programs based on answer set semantics. Alessandra Palmigiano discussed the dynamic nature of categories and their role in human cognition, proposing a comprehensive logical theory of categorization for explainable AI and human-machine interaction.

On the other hand, the 10 papers cover diverse topics in AI and logic, including distributed and field knowledge using weighted modal logic, normative reasoning with balancing operators, recommendation logic, legal norms for AI, epistemic injustice, decidability in tense logic, modal logic with counting, semantic gaps in AI communication, and large language models' biases in responsibility attribution. Other work explores LLMs' capabilities in defeasible reasoning, revealing challenges in handling such information. More specifically, Xiaolong Liang and Yì N. Wáng refine the concept of distributed knowledge by incorporating agents' epistemic abilities and introduces field knowledge as its dual, using epistemic logics interpreted through weighted models. Bin Wei introduced a neuro-symbolic approach that combines LLMs and classical symbolic methods to enhance efficiency, accuracy, and interpretability in pre-litigation mediation and legal outcome prediction. Aleks Knoks, Muyun Shao, Leendert van der Torre, Vincent de Wit, and Liuwen Yu extend their formal framework on normative reasoning by introducing numerical balancing operators to quantify and aggregate the weights of reasons. Fenrong Liu, Wei Wang, and Sisi Yang introduce a new logical framework called Recommendation Logic (RL) to analyze and enhance the reasoning processes behind AI-driven recommendation systems, focusing on user preferences and providing efficient model-checking solutions. Yiwei Lu and Zhe Yu investigate the integration of legal ontology with structured argumentation frameworks to adapt norms for AI actions in legal contexts, utilizing non-monotonic reasoning to reflect dynamic shifts in legal principles. Joris Hulstijn, Huimin Dong, and Réka Markovich explore various forms of epistemic injustice in automated decision-making systems and propose a formalization using epistemic logic to address knowledge discrepancies among different groups. Zhe Yu, Yiheng Wang and Zhe Lin address the decidability of Horn sequent in intuitionistic tense logic S4 by establishing the finite model property, contributing significantly to the resolution of an open problem in the field. Xiaoxuan Fu and Zhiguang Zhao explore the model theoretic properties of modal logic with counting, $ML(\#)$, demonstrating its unique characteristics in terms of model size, compactness, and interpolation, and proving Halldén completeness for certain fragments. Yueying Chu, Jiaxin Zhang, and Peng Liu investigate whether large language models like GPT-3.5 and GPT-4 display human-like biases when attributing responsibility in scenarios involving automated and conventional vehicles. Zhaoqun Li, Chen Chen, and Beishui Liao explore the capabilities of LLMs in defeasible reasoning within the formal logic framework, revealing challenges in handling defeasible information effectively.

We extend our heartfelt gratitude to the invited speakers for their captivating talks, and to the esteemed authors for their invaluable contributions to the workshop. We are also immensely appreciative of the diligent efforts of the program committee members, including Pietro Baroni, Bruno Bentzen, Christoph Benzmüller, Jinsheng Chen, Weiwei Chen, Mehdi Dastani, Huimin Dong, Tiansi Dong, Valeria de Paiva, Guillermo R. Simari, Matthias Thimm, Emil Weydert, Kaibo Xie, and Zhe Yu, who conducted meticulous reviews of the submissions.

Furthermore, we recognize the exceptional dedication and commitment of the local organizers, Wenting Zhang, Sheng Wei, Xiaotong Fang, and their team, who ensured the smooth and successful execution of the event. Last but not least, we acknowledge the financial support provided by the national key project on "Research on Logics for New Generation Artificial Intelligence" for LNGAI 2024, as well as the funding for the Double First-Class Construction initiative from the Academy of Humanities and Social Sciences and the School of Philosophy at Zhejiang University.

Beishui Liao, Zhejiang University, China
Jun Pang, University of Luxembourg, Luxembourg
Tjitze Rienstra, Maastricht University, Netherlands
June 1, 2024

Contents

New directions and considerations for the BOID

Jan Broersen

Utrecht University

Abstract

20 years ago, the BOID-architecture was put forward as an agent modelling framework for reasoning about the selection of goals on the basis of Beliefs, Obligations, (previous) Intentions and Desires. I will critically review our original proposal and suggest new ways of thinking about the BOID, also in relation to LLMs.

Structural Inference of Dynamical Systems: Recent Development and Future Directions

Jun Pang [1]

Department of Computer Science & Institute for Advanced Studies
University of Luxembourg

Abstract

Dynamical systems are pervasive and critical in understanding phenomena across various domains, from the majestic dance of celestial bodies governed by gravity to the subtle ballet of chemical reactions. In the quest to unravel the complexities of dynamical systems, the initial imperative is to unveil their inherent structure, a key determinant of system organisation. Achieving this necessitates the deployment of structural inference methods capable of deriving the structure of dynamical systems from their observed behaviours. In this talk, I will give an overview on recent development of deep learning based methods for structural inference of dynamical systems [1,5], in particular those methods based on variational auto-encoder (VAE) [2,3,6,7,9,4,8]. Through a comprehensive benchmarking study [10], I will also present some key findings and discuss future research directions.

References

[1] Brugere, I., B. Gallagher and T. Y. Berger-Wolf, *Network structure inference, A survey: Motivations, methods, and applications*, ACM Computing Surveys **51** (2018), pp. 24:1–24:39.

[2] Kipf, T., E. Fetaya, K.-C. Wang, M. Welling and R. Zemel, *Neural relational inference for interacting systems*, in: *Proceedings of the 35th International Conference on Machine Learning (ICML)*, PMLR, 2018, pp. 2688–2697.

[3] Löwe, S., D. Madras, R. Z. Shilling and M. Welling, *Amortized causal discovery: Learning to infer causal graphs from time-series data*, in: *Proceedings of the 1st Conference on Causal Learning and Reasoning (CLeaR)* (2022), pp. 509–525.

[4] Pan, L., C. Shi and I. Dokmanić, *A dynamical graph prior for relational inference*, arXiv preprint arXiv:2306.06041 (2023).

[5] Vowels, M. J., N. C. Camgöz and R. Bowden, *D'ya like dags? A survey on structure learning and causal discovery*, ACM Computing Surveys **55** (2023), pp. 82:1–82:36.

[6] Wang, A. and J. Pang, *Iterative structural inference of directed graphs*, in: *Advances in Neural Information Processing Systems 35 (NeurIPS)*, 2022.

[7] Wang, A. and J. Pang, *Active learning based structural inference*, in: *Proceedings of the 40th International Conference on Machine Learning (ICML)* (2023), pp. 36224–36245.

---

[1] fistname.lastname@uni.lu

[8] Wang, A. and J. Pang, *Structural inference with dynamics encoding and partial correlation coefficients*, in: *Proceedings of the 12th International Conference on Learning Representations (ICLR)* (2024).

[9] Wang, A., T. P. Tong and J. Pang, *Effective and efficient structural inference with reservoir computing*, in: *Proceedings of the 40th International Conference on Machine Learning (ICML)* (2023), pp. 36391–36410.

[10] Wang, A., T. P. Tong, J. Pang and A. Mizera, *Benchmarking structural inference methods for dynamical interacting systems*, https://structinfer.github.io/ (2023).

Causality and Responsibility in Multi-agent Systems

Mehdi Dastani

Utrecht University

Abstract

Causality and responsibility are intertwined concepts that play an important role in the reasoning of human and artificial intelligent systems in interactive multi-agent environments. These concepts have been extensively studied in the literature, resulting in a plethora of views and interpretations of these concepts and their relationships. In this presentation, I will introduce a particular view where a group of individuals is held responsible for an outcome if they caused the outcome while they had a strategy to prevent the outcome. To formally instantiate this view, I will propose a systematic approach to modeling interactions in a multi-agent environment based on a given structural causal model. The generated multi-agent model is then used to analyze and reason about the causal effects of agents' strategic decisions and their responsibility.

Human norms, machine norms and AI value alignment

Marija Slavkovik

University of Bergen

Abstract

The talks will consider the problem of AI value alignment from the perspective of the problem of how to automate moral reasoning and decision making for autonomous systems. Normative reasoning has been a sub-discipline in multi agent systems research for a few decades. How does that fit in the age of LLMs? We will situate the normative reasoning work in the larger problem of AI alignment and machine ethics, by discussing the pertinent differences between how value alignment sees norms, why value alignment needs norms and overall what is the role of logic in the world of deep data processing.

Weakest Link, Prioritised Default Logic and Principles in Argumentation

Leendert van der Torre

University of Luxembourg

Abstract

In this article, we study procedural and declarative logics for defaults in modular orders. Brewka's prioritised default logic (PDL) and structured argumentation based on weakest link are compared to each other in different variants. This comparison takes place within the framework of attack relation assignments and the axioms (principles) recently proposed for them by Dung. To this end, we study which principles are satisfied by weakest link and disjoint weakest link attacks. With the aim of approximating PDL using argumentation, we identify an attack defined from PDL extensions,prove that each such PDL extension is a stable belief set under it, and offer a similar principle-based analysis. We also prove an impossibility theorem for Dung's axioms that covers PDL-inspired attack relation assignments. Finally,a novel variant of PDL with concurrent selection of defaults is also proposed,and compared to these argumentative approaches. In sum, our contributions fill an important gap in the literature created by Dung's recent methods and open up new research questions on these methods.

Witnesses for Answer Sets of Logic Programs

Yisong Wang [1]

Guizhou University, Guiyang, Guizhou, China

Thomas Eiter

Institute of Logic and Computation, Technische Universität Wien, Austria

Yuanlin Zhang

Texas Tech University, USA

Fangzhen Lin

Hong Kong University of Science and Technology, Clear Water Bay, Hong Kong

Abstract

Answer Set Programming (ASP) is a declarative problem solving paradigm that can be used to encode a problem as a logic program whose answer sets correspond to the solutions of the problem. It has been widely applied in various domains in AI and beyond. Given that answer sets are supposed to yield solutions to the original problem, the question of "why a set of atoms is an answer set" becomes important for both semantics understanding and program debugging. It has been well investigated for normal logic programs. However, for the class of disjunctive logic programs, which is a substantial extension of that of normal logic programs, this question has not been addressed much. In this talk, we propose a notion of reduct for disjunctive logic programs and show how it can provide answers to the aforementioned question. First, we show that for each answer set, its reduct provides a resolution proof for each atom in it. We then further consider minimal sets of rules that will be sufficient to provide resolution proofs for sets of atoms. Such sets of rules will be called witnesses and are the focus of this article. We study complexity issues of computing various witnesses and provide algorithms for computing them. In particular, we show that the problem is tractable for normal and headcycle-free disjunctive logic programs, but intractable for general disjunctive logic programs. We also conducted some experiments and found that for many well-known ASP and SAT benchmarks, computing a minimal witness for an atom of an answer set is often feasible. These results have been published in ACM Transactions on Computational Logic volume 24 (2) in 2023.

Keywords: Logic programming, minimal models, answer set semantics, witness

[1] yswang@gzu.edu.cn

Categories & categorization

Alessandra Palmigiano

Vrije Universiteit Amsterdam

Abstract

Categories are cognitive tools that humans use to organize their experience, understand and function in the world, and understand and interact with each other, by grouping things together which can be meaningfully compared and evaluated. They are key to the use of language, the construction of knowledge and identity, and the formation of agents' evaluations and decisions. Categorization is the basic operation humans perform e.g. when they relate experiences/actions/objects in the present to ones in the past, thereby recognizing them as instances of the same type. This is what we do when we try and understand what an object is or does, or what a situation means, and when we make judgments or decisions based on experience. The literature on categorization is expanding rapidly in fields ranging from cognitive linguistics to social and management science to AI, and the emerging insights common to these disciplines concern the dynamic essence of categories, and the tight interconnection between the dynamics of categories and processes of social interaction. However, these key aspects are precisely those that both the extant foundational views on categorization struggle the most to address.

In this talk, I will discuss by way of examples how methods, insights, and techniques pertaining to structural proof theory, algebraic logic, duality theory, and category theory in mathematics can be used in synergy with one another to develop an overarching logical theory of categories and categorization, on which new generation explainable AI can be based, as well as a principled approach to human-machine interaction.

Field Knowledge as a Dual to Distributed Knowledge

A Characterization by Weighted Modal Logic

Xiaolong Liang

School of Philosophy
Shanxi University
92 Wucheng Road, Taiyuan, 030006, Shanxi, P.R. China

Yì N. Wáng [1]

Department of Philosophy (Zhuhai)
Sun Yat-sen University
2 Daxue Road, Zhuhai, 519082, Guangdong, P.R. China

Abstract

The study of group knowledge concepts such as mutual, common, and distributed knowledge is well established within the discipline of epistemic logic. In this work, we incorporate epistemic abilities of agents to refine the formal definition of distributed knowledge and introduce a formal characterization of field knowledge. We propose that field knowledge serves as a dual to distributed knowledge. Our approach utilizes epistemic logics with various group knowledge constructs, interpreted through weighted models. We delve into the eight logics that stem from these considerations, explore their relative expressivity and develop sound and complete axiomatic systems.

Keywords: Epistemic logic, weighted model, epistemic skills, distributed knowledge, field knowledge.

1 Introduction

The introduction is segmented into two sections. In Section 1.1, we elucidate our interpretation of the concepts *distributed knowledge* and *field knowledge*. Section 1.2 is dedicated to detailing our methodology for modeling these concepts within the context of weighted (or labeled) modal logic.

---

[1] Corresponding author. Email address: `ynw@xixilogic.org`. The author acknowledges funding support by the National Social Science Fund of China (Grant No. 20&ZD047).

1.1 Group notions of knowledge

Alice and Bob, both instrumentalists with additional expertise in philosophy and mathematics respectively, engage in a conversation that shapes and reflects their knowledge. Classical epistemic logic [12,8,16] offers a framework for dissecting their individual and collective knowledge, utilizing tools like Kripke semantics among others.

Their mutual knowledge (a.k.a. everyone's knowledge or general knowledge) consists of statements known by both, essentially an intersection of their individual knowledge. In Kripke semantics, this is interpreted by the *union* of their respective epistemic uncertainty relations. Common knowledge is recursive mutual knowledge: they know φ, know that they know φ, and so on, ad infinitum. It is modeled by the transitive closure of the union of their uncertainty relations.

Distributed knowledge expresses the sum of knowledge Alice and Bob would have after full communication, but it is not merely the union of what each knows. Though an interpretation based on the *intersection* of their individual uncertainty relations does not fully align with its intended meaning either [18,3], as it stands, this prevalent definition treats *mutual knowledge as the semantic dual to distributed knowledge*.

In our scenario, we conceptualize distributed knowledge in light of the professional competencies of Alice and Bob. Their distributed knowledge of a statement φ is not understood as just a matter of aggregate knowledge, but rather the outcome of their combined expertise in musical instruments, philosophy, and mathematics. That is, φ is their distributed knowledge if their collaborative proficiency across these domains enables them to deduce φ. Thus, we redefine *distributed knowledge* as the *union* of Alice and Bob's epistemic abilities, diverging from its classical interpretation.

Upon reevaluating distributed knowledge, we introduce the allied concept of *field knowledge*. This notion encapsulates knowledge that stems from their shared discipline – musical instruments, in this case. A statement φ falls under Alice and Bob's field knowledge if it is derivable exclusively from their musical background. The formal interpretation of this concept will be presented in Section 2.1, where we propose that *field knowledge semantically functions as the dual to distributed knowledge*.

Developing a coherent characterization for the emergent concepts of distributed and field knowledge presents its challenges within classical epistemic logics. We aim to craft a comprehensive framework that encompasses these new ideas while preserving the established interpretations of mutual and common knowledge.

1.2 Modeling knowledge in weighted modal logic

Even though the concept of *similarity* is intrinsically linked to *knowledge*, it has not been traditionally emphasized or explicitly incorporated in the classical representation of knowledge within the field of epistemic logic [12,8,16]. Over recent years, researchers have started to probe this relationship more deeply, marking a fresh direction in the field [17,7]. The technical framework for exploring this relationship has its roots in weighted modal logics [14,13,11]. This approach offers a quantitative way of considering similarity, allowing for a more nuanced understanding of knowledge.

In this paper, we adapt the concept of similarity from the field of data mining, where it is primarily used to quantify the likeness between two data objects. In data mining, distance and similarity measures are generally specific algorithms tailored to particular scenarios, such as computing the distance and similarity between matrices, texts, graphs, etc. (see, e.g., [1, Chapter 3]). There is also a body of literature that outlines general properties of distance and similarity measures. For instance, in [20], it is suggested that typically, the properties of *positivity* (i.e., $\forall x \forall y : s(x,y) = 1 \Rightarrow x = y$) and *symmetry* (i.e., $\forall x \forall y : s(x,y) = s(y,x)$) hold for $s(x,y)$ – a binary numerical function that maps the similarity between points x and y to the range $[0,1]$.

Our primary interest here does not lie in the measures of similarity themselves, but rather in modeling similarity and deriving from it the concepts of knowledge. Our work distinguishes itself from recent advancements in epistemic logic interpreted through the concepts of similarity or distance [17,7]. One key difference is that we employ the standard language of epistemic logic. We do not explicitly factor in the degree of similarity into the language, maintaining the traditional structure of epistemic logic while reinterpreting its concepts in the light of similarity.

In our setting, the phrase "a knows φ" ($K_a \varphi$) can be interpreted as "φ holds true in all states that, in a's perception by its expertise, resemble the actual state." A "state" in this context can be seen as a data object – the focus of data mining. But it could also be treated as an epistemic object, a possible situation, and so forth. We generalize the similarity function by replacing its range $[0, 1]$ with an arbitrary set of epistemic abilities. The degrees of similarity may not have a comparable or ordered relationship.

The primary focus of this paper is on *group knowledge*, as elaborated in Section 1.1. We explore epistemic logics across all combinations of these group knowledge notions. As mutual knowledge is definable by individual knowledge (with only finitely many agents), we have formulated eight logics (with or without common, distributed, and field knowledge). The syntax and semantics of them are introduced in Sections 2.1–2.2, and in Section 2.3, we compare the expressive power of these languages.

For the axiomatization of the logics, we introduce sound and strongly complete axiomatic systems for the logics excluding common knowledge. For those incorporating common knowledge, we present sound and weakly complete axiomatic systems (owing to the lack of compactness for the common knowledge operators). These systems are then categorized based on whether their completeness results are obtainable via translation of models (Section 3.2.1) or the canonical model method (Section 3.2.2), shown via a path-based canonical model (Sections 3.2.3 and 3.2.4), or require a finitary method leading to a weak completeness result (Section 3.2.5).

2 Logics

In this section, we present a comprehensive framework composed of eight distinctive logics. We supplement our discussion with illustrative examples, offering a visual representation of the models and their accompanying semantics.

2.1 Syntax

Our study utilizes formal languages rooted in the standard language of multi-agent epistemic logic [8,16], with the addition of modalities that represent group knowledge constructs. We particularly concentrate on the constructs of *common knowledge*, *distributed knowledge* and *field knowledge*.

In terms of our assumptions, we consider Prop as a countably infinite set of propositional variables, and Ag as a finite, nonempty set of agents.

Definition 2.1 (formal languages) The languages utilized in our study are defined by the following rules, where the name of each language is indicated in parentheses on the left-hand side:

$$
\begin{array}{ll}
(\mathcal{EL}) & \varphi ::= p \mid \neg\varphi \mid (\varphi \to \varphi) \mid K_a\varphi \\
(\mathcal{ELC}) & \varphi ::= p \mid \neg\varphi \mid (\varphi \to \varphi) \mid K_a\varphi \mid C_G\varphi \\
(\mathcal{ELD}) & \varphi ::= p \mid \neg\varphi \mid (\varphi \to \varphi) \mid K_a\varphi \mid D_G\varphi \\
(\mathcal{ELF}) & \varphi ::= p \mid \neg\varphi \mid (\varphi \to \varphi) \mid K_a\varphi \mid F_G\varphi \\
(\mathcal{ELCD}) & \varphi ::= p \mid \neg\varphi \mid (\varphi \to \varphi) \mid K_a\varphi \mid C_G\varphi \mid D_G\varphi \\
(\mathcal{ELCF}) & \varphi ::= p \mid \neg\varphi \mid (\varphi \to \varphi) \mid K_a\varphi \mid C_G\varphi \mid F_G\varphi \\
(\mathcal{ELDF}) & \varphi ::= p \mid \neg\varphi \mid (\varphi \to \varphi) \mid K_a\varphi \mid D_G\varphi \mid F_G\varphi \\
(\mathcal{ELCDF}) & \varphi ::= p \mid \neg\varphi \mid (\varphi \to \varphi) \mid K_a\varphi \mid C_G\varphi \mid D_G\varphi \mid F_G\varphi
\end{array}
$$

where $p \in$ Prop, $a \in$ Ag, and G represents a nonempty subset of Ag, signifying a group. We also employ other boolean connectives, including conjunction ($\wedge$), disjunction ($\vee$), and equivalence ($\leftrightarrow$). $E_G\varphi$ is a shorthand for $\bigwedge_{a \in G} K_a\varphi$ (note that G is finite).

We employ formulas such as $K_a\varphi$ to depict: "Agent a knows φ." This is often referred to as *individual knowledge*. Similarly, formulas like $C_G\varphi$, $D_G\varphi$, $E_G\varphi$ and $F_G\varphi$ are used to convey that φ is *common knowledge*, *distributed knowledge*, *mutual knowledge* (or *everyone's knowledge*) and field knowledge of group G, respectively. When the group G is a small set, e.g., $\{a, b\}$, we write $C_{ab}\varphi$ as a shorthand for $C_{\{a,b\}}\varphi$, and likewise for the operators D_{ab}, E_{ab} and F_{ab}.

Before delving into the formal semantics of these formulas, it is important to first establish the semantic models that will be used for the intended logics.

2.2 Semantics

We introduce a type of *similarity models* for the interpretation of the languages.

Definition 2.2 (similarity models) A *similarity model* (*model* for short) is a quintuple (W, A, E, C, ν) where:

- W is a nonempty set of states or nodes, referred to as the *domain*;
- A is an arbitrary set of abstract epistemic *abilities* (e.g., one's expertise,

12

skills, professions or privileges), which could be empty, finite or infinite; [2]
- $E : W \times W \to \wp(A)$, known as an *edge function*, assigns each pair of states a set of epistemic abilities, meaning that the two states are indistinguishable for individuals possessing only these epistemic abilities;
- $C : \mathsf{Ag} \to \wp(A)$ is a *capability function* that assigns each agent a set of epistemic abilities;
- $\nu : W \to \wp(\mathsf{Prop})$ is a valuation.

and conforms to the following conditions (for all $s, t \in W$):
- Positivity: if $E(s, t) = A$, then $s = t$;
- Symmetry: $E(s, t) = E(t, s)$.

The above definition requires further elucidation. Firstly, our approach adopts a broad interpretation of epistemic abilities that may not necessarily be arranged in a linear order, although such an arrangement is plausible [7,15]. Secondly, we perceive the edge function E as a representation of the relation of similarity between states. The conditions of positivity and symmetry serve as generalized forms of common conditions employed to characterize similarity between data objects, as demonstrated in [20]. [3] Transitivity is usually not a characteristic of this framework (that x and y, y and z are similar, does not necessarily mean that x and z are similar), resulting in the failure to uphold the principle of positive introspection ($K_a\varphi \to K_a K_a\varphi$). Although it is easy to impose transitivity, we choose not to enforce it here. Our framework allows a more discerning evaluation of the tenability of positive introspection, and for an examination of other significant constraints, see [15]. Thirdly, in this context, similarities are deemed objective, signifying their constancy across diverse agents.

Example 2.3 Alice and Bob are denoted as a and b, respectively. Consider the fields mentioned in the beginning of the paper: musical instruments (α), philosophy (β) and mathematics (γ), which are regarded as epistemic abilities in this example. As set up in the beginning of the paper, Alice is a philosopher, Bob a mathematician, and both of them are also instrumentalists. Four possible states are named $s_1, \ldots, s_4$, in which s_1 is the factual state. From the viewpoint of an instrumentalist, all the states look no difference. From the perspective of a philosopher, no difference is between s_1 and s_3, and between s_2 and s_4. As for a mathematician, s_1 is indistinguishable to s_2, and s_3 to s_4. Consider the

---

[2] We have opted not to fix the set A of abilities as a given parameter of the logic, in contrast to the set Ag of agents. The primary reason for this decision is our intention to examine models that may extend the set A (see Section 3.2.1). It is important to note that the validities and subsequent axiomatization of our logic remain unaffected when A is considered to be an infinite set of abilities defined as a parameter of the logic.

[3] An implicit condition often assumed, the converse of positivity, posits $E(s, t) = A$ if $s = t$. This condition entails the reflexivity of graphs, depicted by the characterization axiom T (i.e., $K_a\varphi \to \varphi$). In the realm of data mining, this condition implies that if two data objects are identical (i.e., they refer to the same data object), they would receive the maximum value from any similarity measure. This, however, is not always guaranteed.

following propositions:

p: A standard modern piano has 88 keys in total.

q: Knowledge is defined by "justified true belief."

r: Fermat's Last Theorem has been proved.

The above scenario can be abstracted to a pointed model (M, s_1), such that $M = (W, A, E, C, \nu)$ and:

- $W = \{s_1, s_2, s_3, s_4\}$; $A = \{\alpha, \beta, \gamma\}$;
- $E(s_1, s_2) = E(s_3, s_4) = \{\alpha, \gamma\}$, $E(s_1, s_3) = E(s_2, s_4) = \{\alpha, \beta\}$, $E(s_1, s_4) = E(s_2, s_3) = \{\alpha\}$, and for all $x \in W$, $E(x, x) = \{\alpha, \beta, \gamma\}$;
- $C(a) = \{\alpha, \beta\}$ and $C(b) = \{\alpha, \gamma\}$;
- $\nu(s_1) = \{p, r\}$, $\nu(s_2) = \{p, q, r\}$, $\nu(s_3) = \{p\}$ and $\nu(s_4) = \{p, q\}$.

Figure 1 illustrates the model M introduced above, where the factual state s_1 is framed by a rectangle and other states by an eclipse.

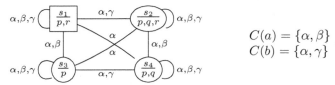

$$C(a) = \{\alpha, \beta\}$$
$$C(b) = \{\alpha, \gamma\}$$

Fig. 1. Illustration of the model in Example 2.3.

In the real world (s_1), one may come up with the following true sentences:

- Alice knows p and $\neg q$, but doesn't know whether r.
  To be formulated by: $K_a(p \wedge \neg q) \wedge \neg(K_a r \vee K_a \neg r)$.
- Bob knows that p and r, but doesn't know whether q.
  To be formulated by: $K_b(p \wedge r) \wedge \neg(K_a q \vee K_a \neg q)$.
- It is Alice and Bob's distributed knowledge that p, not q, and r.
  To be formulated by: $D_{ab}(p \wedge \neg q \wedge r)$.
- While p is Alice and Bob's field knowledge, q and r are not.
  To be formulated by: $F_{ab}p \wedge \neg(F_{ab}q \vee F_{ab}\neg q) \wedge \neg(F_{ab}r \vee F_{ab}\neg r)$.

We now introduce a formal semantics that makes the model in Example 2.3 indeed yields the true sentences listed above.

Definition 2.4 Given a formula φ, a model $M = (W, A, E, C, \nu)$ and a state $s \in W$, we say φ is *true* or *satisfied* at s in M, denoted $M, s \models \varphi$, if the following hold (the case for $E_G \psi$ is redundant, but included for clarification):

$$
\begin{aligned}
M, s \models p & \iff p \in \nu(s) \\
M, s \models \neg\psi & \iff \text{not } M, s \models \psi \\
M, s \models (\psi \to \chi) & \iff \text{if } M, s \models \psi \text{ then } M, s \models \chi \\
M, s \models K_a \psi & \iff \text{for all } t \in W, \text{ if } C(a) \subseteq E(s, t) \text{ then } M, t \models \psi \\
M, s \models E_G \psi & \iff M, s \models K_a \psi \text{ for all } a \in G \\
M, s \models C_G \psi & \iff \text{for all positive integers } n, M, s \models E_G^n \psi \\
M, s \models D_G \psi & \iff \text{for all } t \in W, \text{ if } \bigcup_{a \in G} C(a) \subseteq E(s, t) \text{ then } M, t \models \psi \\
M, s \models F_G \psi & \iff \text{for all } t \in W, \text{ if } \bigcap_{a \in G} C(a) \subseteq E(s, t) \text{ then } M, t \models \psi,
\end{aligned}
$$

14

where $E_G^n \psi$ is defined recursively as $E_G^1 E_G^{n-1} \psi$, with $E_G^1 \psi$ to mean $E_G \psi$. The concepts of *validity* and *satisfiability* have their classical meaning.

In the definition above, the interpretation of $K_a \psi$ includes a condition "$C(a) \subseteq E(s,t)$," which intuitively means that, "Agent a, with its abilities, cannot discern between states s and t." Thus, the formula $K_a \psi$ expresses that ψ is true in all states t that a cannot differentiate from the current state s.

$E_G \psi$ stands for the conventional notion of *everyone's knowledge*, or *mutual knowledge* as we call, stating that "Everyone in group G knows ψ" (see [10] for details).

Common knowledge ($C_G \psi$) follows the classical fixed-point interpretation as $E_G C_G \psi$. In other words, $C_G \psi$ implies that, "Everyone in group G knows that ψ is true, and everyone in G knows about this first-order knowledge, and also knows about this second-order knowledge, and so on."

The concept of *distributed knowledge* ($D_G \psi$) in this paper diverges from the traditional definitions found in literature. We redefine distributed knowledge as being attainable by pooling together individual abilities. In practice, we swap the intersection of individual uncertainty relations with the union of individual epistemic abilities. Thus, ψ is deemed distributed knowledge among group G if and only if ψ holds true in all states t that, when utilizing all the epistemic abilities of agents in group G, cannot be differentiated from the present state.

An additional type of group knowledge, termed *field knowledge* ($F_G \psi$), states that ψ is field knowledge if and only if ψ is true in all states t that, using the shared abilities of group G, cannot be differentiated from the current state. We will examine its logical properties in greater detail later on.

Upon defining the semantics, we derive eight logics, each associated with one of the languages. We denote these logics using upright Roman capital letters. E.g., the logic corresponding to the interpretation of the language $\mathcal{ELF}$ is represented as ELF.

Example 2.5 Consider the model illustrated in Figure 2, we have the following:

(i) $M, s_2 \models C_{ab}p \wedge E_{ab}p \wedge D_{ab}p \wedge \neg F_{ab}p$ (note that for $C_{ab}p$ we check whether p is true in all states along an "ab-path" – connected via $\{\lambda, \pi\}$, $\{\lambda, \mu\}$ or $\{\lambda, \pi, \mu\}$ edges – that is, whether p is true at s_2 through s_4, regardless of s_1.)

(ii) $M, s_2 \models F_{ab}q \wedge E_{ab}q \wedge D_{ab}q \wedge \neg C_{ab}q$

(iii) $M, s_3 \models D_{ab}r \wedge \neg K_a r \wedge \neg K_b r \wedge \neg E_{ab}r$

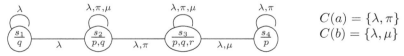

Fig. 2. Illustration of a model for Example 2.5. We do not draw a line between two nodes when the edge between them is with no label (i.e., labeled by an empty set, e.g., between s_1 and s_3).

From the above, it is clear that none of the following formula schemes are valid: $C_G\varphi \to F_G\varphi$, $E_G\varphi \to F_G\varphi$, $D_G\varphi \to F_G\varphi$, $F_G\varphi \to C_G\varphi$, $E_G\varphi \to C_G\varphi$, $D_G\varphi \to C_G\varphi$, $D_G\varphi \to E_G\varphi$, $D_G\varphi \to K_a\varphi$ (where $a \in G$). In particular, that φ is common knowledge implies that φ is distributed knowledge ($\models C_G\varphi \to D_G\varphi$), but does not imply that it is field knowledge ($\not\models C_G\varphi \to F_G\varphi$). The underlying reasoning for this is that when professions intersect, the range of uncertain states can potentially expand significantly – sometimes even more so than the increase that occurs when taking the transitive closure in the case of common knowledge. Consequently, this expansion of uncertainty can lead to a substantial contraction of field knowledge. Nonetheless, the standard principles pertaining to individual, common, and distributed knowledge from classical logic remain applicable, as indicated by the following proposition.

Proposition 2.6 *We have the following validities for any given formula φ, any agent a and any groups G and H (proofs omitted):*

(i) $K_a(\varphi \to \psi) \to (K_a\varphi \to K_a\psi)$

(ii) $\varphi \to K_a\neg K_a\neg\varphi$

(iii) $C_G\varphi \to \bigwedge_{a \in G} K_a(\varphi \wedge C_G\varphi)$

(iv) $D_{\{a\}}\varphi \leftrightarrow K_a\varphi$

(v) $D_G\varphi \to D_H\varphi$ *(with $G \subseteq H$)*

(vi) $\varphi \to D_G\neg D_G\neg\varphi$

(vii) $F_{\{a\}}\varphi \leftrightarrow K_a\varphi$

(viii) $F_G\varphi \to F_H\varphi$ *(with $H \subseteq G$)*

(ix) $\varphi \to F_G\neg F_G\neg\varphi$

2.3 Expressivity

We adopt the conventional method for assessing a language's expressive power, which entails benchmarking it against the expressive capabilities of other languages. For an exact articulation of the relations in expressive power between two compatibly interpreted languages, we direct the reader to [21, Def. 8.2].

It is evident that the expressive power of all eight languages under consideration can be evaluated against one another. The comparative outcomes are encapsulated in Figure 3, and the proofs are left in Appendix A.

3 Axiomatization

We will present sound and complete axiomatic systems for the logics introduced in the preceding section. The names of these systems will be designated with bold capital letters. For example, the axiomatic system for the logic ELF is denoted as **ELF**.

3.1 Axiomatic systems

The system **K** is a widely recognized axiomatic system for modal logic (here it refers to the multi-modal version with each K_a functioning as a box operator). For simplicity, the axiom schemes are referred to as axioms in this context. The axiom system **KB** is obtained by augmenting the system **K** with an additional axiom B (i.e., $\varphi \to K_a\neg K_a\neg\varphi$). In this context, we represent **KB** as $\mathbf{K} \oplus \mathbf{B}$, where the symbol $\oplus$ acts like a union operation for sets of axioms and/or rules. For a comprehensive understanding of these axiomatic systems for modal logic,

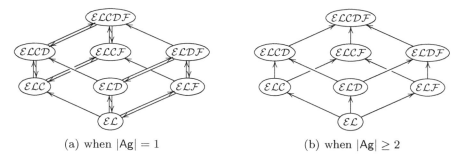

(a) when $|\mathsf{Ag}| = 1$ (b) when $|\mathsf{Ag}| \geq 2$

Fig. 3. The above two diagrams illustrate the relative expressive power of the languages. An arrow pointing from one language to another implies that the second language is at least as expressive as the first. The "at least as expressive as" relationship is presumed to be *reflexive* and *transitive*, meaning that a language is considered at least as expressive as another if a path of arrows exists leading from the second to the first (self-loops exist for all, but omitted). A lack of a path of arrows from one language to another indicates that the first language is *not* at least as expressive as the second. This implies that either the two languages are incomparable or that the first language is more expressive than the second.

please refer to, say, [5]. Now, the system **EL** that we introduce for our base logic EL is in fact **KB**.

Common knowledge is characterized by a set **C** consisting of the following two inductive principles, represented by an axiom and a rule, which can be found in [8]:

(C1) $C_G\varphi \rightarrow \bigwedge_{a \in G} K_a(\varphi \wedge C_G\varphi)$

(C2) from $\varphi \rightarrow \bigwedge_{a \in G} K_a(\varphi \wedge \psi)$ infer $\varphi \rightarrow C_G\psi$

Our the system **ELC** is then represented as **EL** $\oplus$ **C**.

Distributed knowledge is characterized by a set **D** of additional axioms:

(KD) $D_G(\varphi \rightarrow \psi) \rightarrow (D_G\varphi \rightarrow D_G\psi)$

(D1) $D_{\{a\}}\varphi \leftrightarrow K_a\varphi$

(D2) $D_G\varphi \rightarrow D_H\varphi$ with $G \subseteq H$

(BD) $\varphi \rightarrow D_G\neg D_G\neg\varphi$

The resulting system for the logic ELD is then denoted as **ELD** = **EL** $\oplus$ **D**.

Field knowledge is characterized by the following set **F**:

(KF) $F_G(\varphi \rightarrow \psi) \rightarrow (F_G\varphi \rightarrow F_G\psi)$

(F1) $F_{\{a\}}\varphi \leftrightarrow K_a\varphi$

(F2) $F_G\varphi \rightarrow F_H\varphi$ with $H \subseteq G$

(BF) $\varphi \rightarrow F_G\neg F_G\neg\varphi$

(NF) from φ infer $F_G\varphi$

While the set **F** might at first glance seem analogous to **D**, there exists a subtle yet crucial difference between the axioms F2 and D2 – specifically, the positions of the groups G and H are swapped. This distinction necessitates the introduction of the necessitation rule NF, while within the system **ELD**,

17

the rule "from φ infer $D_G\varphi$" is derivable. The validity of these axioms can be verified with relative ease, and it is particularly noteworthy how the altered order of G and H precisely mirrors the union/intersection of epistemic abilities as observed in the semantic interpretations. Similarly, **ELF** is represented as **EL** $\oplus$ **F**.

Moving towards more complex axiomatic systems, they are constructed in a similar manner. For any given string Ξ comprising elements from the set $\{\mathbf{C}, \mathbf{D}, \mathbf{M}\}$:

The axiomatic system **EL**Ξ *consists of all axioms and rules of* **EL***, along with those of the sets denoted by each character in string* Ξ*.*

To illustrate, when Ξ is the string "**CF**," **ELCF** stands for **EL** $\oplus$ **C** $\oplus$ **F**, and when Ξ is the string "**CDF**," then **ELCDF** equates to **EL** $\oplus$ **C** $\oplus$ **D** $\oplus$ **F**. For an extreme case, when Ξ is an empty string, **EL**Ξ simply stands for **EL**.

We now turn our attention to validating these axiomatic systems to be sound and complete for the corresponding logics. Soundness states that all the theorems of an axiomatic system are valid sentences of the corresponding logic. This can be simplified to the task of verifying that all the axioms of the system are valid, and that all the rules preserve this validity. The soundness of the proposed axiomatic systems can be confirmed without much difficulty. Though we omit the proof, we state it as the following theorem. We will follow this up with the completeness results in the subsequent section.

Theorem 3.1 (soundness) *Every axiomatic system introduced in this section is sound for its corresponding logic.* □

3.2 Completeness

In this section, we aim to demonstrate the completeness of all eight axiomatic systems that were introduced earlier.

It is a widely accepted fact in classical epistemic logic that the inclusion of common knowledge can cause a logic to lose its *compactness*. This leads to the situation where its axiomatic system is not strongly complete, but only weakly complete (see, e.g., [5,21]). This is also the case in our context. As a consequence, we will demonstrate that the four systems that do not include common knowledge are strongly complete axiomatic systems for their corresponding logics, while the other four systems that do incorporate common knowledge are only weakly complete.

The structure of this section is predicated on the various proof techniques we employ. We start with a method that reduces the satisfiability from classical epistemic logics to the logics we have proposed (Section 3.2.1). However, this technique is only applicable to the system **EL**; the canonical model method also works, and we provide a definition in Section 3.2.2 for the reference of the reader. When dealing with systems that incorporate either distributed or field knowledge, but not both (i.e., **ELD** and **ELF**), we utilize a path-based canonical model method (Section 3.2.3). For the system that includes both distributed and field knowledge, namely **ELDF**, while the path-based canonical model method still applies, a slightly more nuanced approach is required (Section 3.2.4). Lastly, for the remaining systems incorporating common knowledge, we merge the finitary method (which involves constructing a closure) with the methods mentioned above (Section 3.2.5).

3.2.1 Proof by translation of satisfiability

Definition 3.2 (translation ρ) The mapping $\cdot^\rho$ from symmetric Kripke models to similarity models is such that, for any symmetric Kripke model $N = (W, R, V)$, N^ρ is the similarity model $(W, \mathsf{Ag} \cup \{b\}, E, C, \nu)$ with the same domain and:

- E is such that for all $s, t \in W$, $E(s,t) = \{a \in \mathsf{Ag} \mid (s,t) \in R(a)\}$,
- b is a new agent which does not appear in Ag,
- C is such that for all $a \in \mathsf{Ag}$, $C(a) = \{a\}$, and
- ν is such that for all $s \in W$, $\nu(s) = \{p \in \mathsf{Prop} \mid s \in V(p)\}$.

In the translated model N^ρ of the above definition, the set of epistemic abilities is appointed as $\mathsf{Ag} \cup \{b\}$. We use agents as labels of edges, which can intuitively be understood as an agent's inability to distinguish the ongoing state from current state when considering their epistemic abilities as a whole. In the subsequent lemma, we demonstrate that this translation preserves truth.

Lemma 3.3 *The following hold (proof in Appendix B):*

(i) *Given a symmetric Kripke model N, its translation N^ρ is a similarity model;*

(ii) *For any $\mathcal{ELCD}$-formula φ, any symmetric Kripke model N and any state s of N, $N, s \Vdash \varphi$ iff $N^\rho, s \models \varphi$.* [4]

The system **EL** is known to be complete for symmetric Kripke models. By Lemma 3.3, such a model can be translated to a similarity model in a way that preserves truth. So we get the following.

Theorem 3.4 EL *is strongly complete for* EL. □

We note that it is possible to use the same method to achieve completeness results for the systems **ELC**, **ELD** and **ELCD**, as long as their completeness results in classical epistemic logic exist. However, to the best of our knowledge, the completeness of these systems in Kripke semantics, while expected, has never been explicitly established.

3.2.2 Proof by the canonical model method

We have demonstrated the completeness of **EL** using the method of translation. This method is efficient and relies on the completeness result for its classical counterpart interpreted via Kripke semantics. In other words, the translation method cannot be employed for logics whose classical counterparts have not been introduced or studied (for example, logics with field knowledge), or when

---

[4] The operator F_G is undefined in classical epistemic logic, so there is no point to consider a language incorporated with this operator.

a completeness result does not exist for their classical counterparts (for instance, ELC, ELD and ELCD). In this section, we introduce a direct proof of the completeness of **EL** using the canonical model method and extend this to completeness proofs for other logics in later sections.

Definition 3.5 The *canonical model for* EL is the tuple $M = (W, A, E, C, v)$ such that:

- W is the set of all maximal **EL**-consistent sets of $\mathcal{EL}$-formulas;
- $A = \wp(Ag)$, the power set of all agents;
- $E : W \times W \to \wp(A)$ is defined such that for any $\Phi, \Psi \in W$, $E(\Phi, \Psi) = \bigcup_{a \in Ag} E_a(\Phi, \Psi)$, where

$$E_a(\Phi, \Psi) = \begin{cases} C(a), & \text{if } \{\chi \mid K_a\chi \in \Phi\} \subseteq \Psi \text{ and } \{\chi \mid K_a\chi \in \Psi\} \subseteq \Phi, \\ \emptyset, & \text{otherwise;} \end{cases}$$

- $C : Ag \to \wp(A)$ is such that for any agent a, $C(a) = \{G \subseteq Ag \mid a \in G\}$;
- $v : W \to \wp(Prop)$ is such that for any $\Phi \in W$, $v(\Phi) = \{p \in Prop \mid p \in \Phi\}$.

The discerning reader may first ascertain that the canonical model for EL is indeed a model, and then proceed to demonstrate the completeness of **EL** by employing conventional techniques. The specifics of these procedures are left in Appendix C.

3.2.3 Proof using a path-based canonical model

When addressing logics that incorporate concepts of distributed and field knowledge, the conventional canonical model technique is not suitable. To overcome this challenge, there exists a methodology for logics with distributed knowledge [9]. This technique, which traces its origins to the unraveling methods of [19], starts by analogously treating distributed knowledge as a form of individual knowledge. The process involves creating a pseudo model that embodies these aspects. This pseudo model is then unraveled into a tree-like structure with paths. The resulting structure is further processed through an identification/folding step to yield the target model.

An simplified approach has been proposed in [22], where the construction of a path-based, tree-like model, referred to as a *standard model*, is advocated. This approach eliminates the intermediate steps of unraveling and identification/folding. We embrace this latter approach here, setting out to construct a standard model directly.

Completeness of ELD

Definition 3.6 $\langle \Phi_0, G_1, \Phi_1, \ldots, G_n, \Phi_n \rangle$ is called a *canonical path* for ELD, if:

- $\Phi_0, \Phi_1, \ldots, \Phi_n$ represent maximal **ELD**-consistent sets of $\mathcal{ELD}$-formulas,
- $G_1, \ldots, G_n$ denote groups of agents, i.e., nonempty subsets of Ag.

In the context of a canonical path $s = \langle \Phi_0, G_1, \Phi_1, \ldots, G_n, \Phi_n \rangle$, we denote Φ_n as $tail(s)$. This also applies to canonical paths defined later.

Definition 3.7 The *standard model for* ELD is the tuple $M = (W, A, E, C, v)$ such that:

- W is the set of all canonical paths for ELD;
- A = $\wp(\mathsf{Ag})$;
- E : W × W → $\wp(\mathsf{A})$ is defined such that for any $s, t \in$ W, E$(s, t) =$

$$
\begin{cases}
\bigcup_{a \in G} \mathsf{C}(a), & \text{if } t \text{ is } s \text{ extended with } \langle G, \Psi \rangle, \\
& \{\chi \mid D_G\chi \in tail(s)\} \subseteq \Psi \text{ and } \{\chi \mid D_G\chi \in \Psi\} \subseteq tail(s); \\
\bigcup_{a \in G} \mathsf{C}(a), & \text{if } s \text{ is } t \text{ extended with } \langle G, \Psi \rangle, \\
& \{\chi \mid D_G\chi \in tail(t)\} \subseteq \Psi \text{ and } \{\chi \mid D_G\chi \in \Psi\} \subseteq tail(t); \\
\emptyset, & \text{otherwise};
\end{cases}
$$

- C : $\mathsf{Ag} \to \wp(\mathsf{A})$ is such that for any agent a, C$(a) = \{G \in \mathsf{Ag} \mid a \in G\}$;
- v : W → $\wp(\mathsf{Prop})$ is such that for all $s \in$ W, v$(s) = \{p \in \mathsf{Prop} \mid p \in tail(s)\}$.

Lemma 3.8 (standardness) *The standard model for* ELD *is a model.*

Proof. Note that $\emptyset \notin$ C(a) for any agent a. This implies that for any $s, t \in$ W, E$(s, t) \neq$ A, thereby meeting the criterion of positivity. Additionally, the condition of symmetry is fulfilled as E is a commutative function. □

Lemma 3.9 (Truth Lemma) *Let* M $= ($W, A, E, C, v$)$ *be the standard model for* ELD. *For any* $s \in$ W *and* $\mathcal{ELD}$-*formula* φ, $\varphi \in tail(s)$ *iff* M$, s \models_{\text{ELD}} \varphi$.

Proof. We prove it by induction of φ. The boolean cases are easy by the definition of ν and the induction hypothesis. The only interested cases are $\varphi = K_a\psi$ and $\varphi = D_G\psi$.

Case $\varphi = K_a\psi$: very similar to the case for $D_G\psi$ given below.

Case $\varphi = D_G\psi$: Suppose $D_G\psi \notin tail(s)$, but M$, s \models_{\text{ELD}} D_G\psi$. By definition, $\bigcup_{a \in G}$ C$(a) \subseteq$ E(s, t) implies M$, t \models_{\text{ELD}} \psi$ for any $t \in$ W. We can extend $\{\neg\psi\} \cup \{\chi \mid D_G\chi \in tail(s)\} \cup \{\neg D_G\neg\chi \mid \chi \in tail(s)\}$ to some maximal **ELD**-consistent set Δ^+. Similar to the proof of Lemma C.2 we get $\bigcup_{a \in G}$ C$(a) \subseteq$ E(s, t) where t extends s with $\langle G, \Delta^+\rangle$. By the induction hypothesis we have M$, t \models_{\text{ELD}} \neg\psi$. A contradiction!

For the opposite direction, suppose $D_G\psi \in tail(s)$, but M$, s \not\models_{\text{ELD}} D_G\psi$. Then there exists $t \in W$ such that M$, t \not\models_{\text{ELD}} \psi$ and $\bigcup_{a \in G} C(a) \subseteq$ E(s, t). This implies that there exists $H \supseteq G$, such that $\{\chi \mid D_H\chi \in tail(s)\} \subseteq tail(t)$ and $\{\chi \mid D_H\chi \in tail(t)\} \subseteq tail(s)$. Since $D_G\psi \in tail(s)$ implies $D_H\psi \in tail(s)$, we have $\psi \in tail(t)$. By the induction hypothesis we have M$, t \models_{\text{ELD}} \psi$, also leading to a contradiction. □

Theorem 3.10 ELD *is strongly complete for* ELD. □

Completeness of ELF The completeness of **ELF** can be demonstrated in a manner that parallels the completeness of **ELD**. While we will not delve into the intricate details of the proofs, we will outline the necessary adaptations to the definitions of the standard model.

A canonical path for ELF mirrors that for ELD. The only modification required is the adjustment of the maximal consistent sets to align with the axiomatic system being considered. When defining the standard model for

ELF, we replace W with the set of all canonical paths for ELF, and let

$$\mathsf{E}(s,t) = \begin{cases} \bigcap_{a \in G} \mathsf{C}(a), & \text{if } t \text{ is } s \text{ extended with } \langle G, \Psi \rangle, \\ & \{\chi \mid F_G\chi \in tail(s)\} \subseteq \Psi \text{ and } \{\chi \mid F_G\chi \in \Psi\} \subseteq tail(s), \\ \bigcap_{a \in G} \mathsf{C}(a), & \text{if } s \text{ is } t \text{ extended with } \langle G, \Psi \rangle, \\ & \{\chi \mid F_G\chi \in tail(t)\} \subseteq \Psi \text{ and } \{\chi \mid F_G\chi \in \Psi\} \subseteq tail(t), \\ \emptyset, & \text{otherwise.} \end{cases}$$

Note that $\bigcup_{a \in G} \mathsf{C}(a) = \{H \mid H \cap G \neq \emptyset\}$, which includes all the shared epistemic abilities of those groups H that intersects with G. Additionally, $\bigcap_{a \in G} \mathsf{C}(a) = \{H \mid G \subseteq H\}$, which represents all the supersets of G.

By using analogous proof structures, we can demonstrate the standardness of these models, show a Truth Lemma, and establish the completeness.

Theorem 3.11 ELF *is strongly complete for* ELF. $\qquad\qquad\square$

3.2.4 Incorporation of both distributed and field knowledge

We now discuss the logic and its axiomatic system that incorporate both distributed and field knowledge but exclude common knowledge, namely the logic ELDF and the axiomatic system **ELDF**. The construction process requires careful consideration of the intricate interaction between the two types of knowledge modalities.

Definition 3.12 $\langle \Phi_0, I_1, \Phi_1, \ldots, I_n, \Phi_n \rangle$ is a *canonical path for* ELDF, if:

- $\Phi_0, \Phi_1, \ldots, \Phi_n$ are maximal **ELDF**-consistent sets of $\mathcal{ELDF}$-formulas;
- $I_1, \ldots, I_n$ are of the form (G, d) or (G, m), with G denoting a group, and "d" and "m" being just two distinct characters.

Definition 3.13 The *standard model for* ELDF is a tuple $\mathsf{M} = (\mathsf{W}, \mathsf{A}, \mathsf{E}, \mathsf{C}, \mathsf{v})$ where A, C and v are defined just as in the standard model for ELD (Definition 3.7), and:

- W is the set of all canonical paths for ELDF;
- $\mathsf{E} : \mathsf{W} \times \mathsf{W} \to \wp(\mathsf{A})$ is such that for any $s, t \in \mathsf{W}$, $\mathsf{E}(s,t) =$

$$\begin{cases} \bigcup_{a \in G} \mathsf{C}(a), & \text{if } t \text{ extends } s \text{ with } \langle (G, d), \Psi \rangle, \\ & \{\chi \mid D_G\chi \in tail(s)\} \subseteq \Psi \text{ and } \{\chi \mid D_G\chi \in \Psi\} \subseteq tail(s), \\ \bigcup_{a \in G} \mathsf{C}(a), & \text{if } s \text{ extends } t \text{ with } \langle (G, d), \Psi \rangle, \\ & \{\chi \mid D_G\chi \in tail(t)\} \subseteq \Psi \text{ and } \{\chi \mid D_G\chi \in \Psi\} \subseteq tail(t), \\ \bigcap_{a \in G} \mathsf{C}(a), & \text{if } t \text{ extends } s \text{ with } \langle (G, m), \Psi \rangle, \\ & \{\chi \mid F_G\chi \in tail(s)\} \subseteq \Psi \text{ and } \{\chi \mid F_G\chi \in \Psi\} \subseteq tail(s), \\ \bigcap_{a \in G} \mathsf{C}(a), & \text{if } s \text{ extends } t \text{ with } \langle (G, m), \Psi \rangle, \\ & \{\chi \mid F_G\chi \in tail(t)\} \subseteq \Psi \text{ and } \{\chi \mid F_G\chi \in \Psi\} \subseteq tail(t), \\ \emptyset, & \text{otherwise.} \end{cases}$$

The standardness and the Truth Lemma can be achieved in a similar way, and we leave a proof of the Truth Lemma in Appendix D for the careful reader.

Theorem 3.14 ELDF *is strongly complete for* ELDF. $\qquad\qquad\square$

3.2.5 Proof by a finitary standard model

We now delineate the extension of the completeness results to the rest of the logics with common knowledge, deploying a finitary method for this purpose. We can only achieve weak completeness due to the non-compact nature of the common knowledge modality. A difficulty, except for ELC, is that we also need to address the modality for distributed or field knowledge.

In this section, we focus on providing the completeness proofs for **ELCDF**. By making simple adaptations, one can obtain the completeness of the axiomatic systems for their sublogics with common knowledge. We adapt the definition of the *closure* of a formula presented in [22], to cater to formulas with modalities D_G and/or F_G.

Definition 3.15 For an $\mathcal{ELCDF}$-formula φ, we define $cl(\varphi)$ as the minimal set satisfying the subsequent conditions:

(i) $\varphi \in cl(\varphi)$;

(ii) if ψ is in $cl(\varphi)$, so are all subformulas of ψ;

(iii) $\psi \in cl(\varphi)$ implies $\sim\psi \in cl(\varphi)$, where $\sim\psi = \neg\psi$ if ψ is not a negation and $\sim\psi = \chi$ if $\psi = \neg\chi$;

(iv) $K_a\psi \in cl(\varphi)$ implies $D_{\{a\}}\psi, F_{\{a\}}\psi \in cl(\varphi)$;

(v) $D_{\{a\}}\psi \in cl(\varphi)$ implies $K_a\psi \in cl(\varphi)$;

(vi) For groups G and H, if H appears in φ, then $D_G\psi \in cl(\varphi)$ implies $D_H\psi \in cl(\varphi)$;

(vii) $C_G\psi \in cl(\varphi)$ implies $\{K_a\psi, K_aC_G\psi \mid a \in G\} \subseteq cl(\varphi)$;

(viii) $F_G\psi \in cl(\varphi)$ implies $\{K_a\psi \mid a \in G\} \subseteq cl(\varphi)$;

(ix) For groups G and H, if H appears in φ, then $F_G\psi \in cl(\varphi)$ implies $F_H\psi \in cl(\varphi)$.

Considering that the initial two clauses exclusively introduce subformulas of φ, the subsequent three clauses incorporate formulas in a constrained manner, and given that there are a finite number of groups mentioned in φ, with each group containing only a finite number of agents, we can readily confirm that $cl(\varphi)$ for any given formula φ.

Subsequently, we introduce the concept of a *maximal consistent set of formulas within a closure*. For a comprehensive definition, which is naturally contingent on the specific axiomatic system under consideration, we refer to established literature, for example, [21].

A *canonical path for* ELCDF *in* $cl(\varphi)$ is defined similarly to that for ELDF (Def. 3.12). Given an $\mathcal{ELCDF}$-formula φ, we can construct the *standard model for* ELCDF *with respect to* $cl(\varphi)$ in a manner that closely mirrors the construction of the standard model for ELDF (as per Def. 3.13). The primary differences lie in bounding the canonical paths by the closure and adjusting the logics accordingly. More specifically, we need to: (1) replace all occurrences of "$\mathcal{ELDF}$" with "$\mathcal{ELCDF}$," and "ELDF" with "ELCDF"; (2) within the definition of W, replace "canonical paths for ELDF" with "canonical paths for ELCDF in $cl(\varphi)$." Moreover, it is easy to confirm that the standard model for

ELCDF (in any closure of a given formula) is a model.

Lemma 3.16 (Truth Lemma) *Given an $\mathcal{ELCDF}$-formula θ, and let* $\mathsf{M} = (\mathsf{W}, \mathsf{A}, \mathsf{E}, \mathsf{C}, \mathsf{v})$ *be the standard model for* ELCDF *with respect to* $cl(\theta)$, *for any* $s \in \mathsf{W}$ *and* $\varphi \in cl(\theta)$, *we have* $\varphi \in tail(s)$ *iff* $\mathsf{M}, s \models_{\mathrm{ELCDF}} \varphi$.

Proof. We show the lemma by induction on φ. We omit the straightforward cases (partially found in Appendix E) and focus on the cases concerning common knowledge.

Suppose $C_G\psi \in tail(s)$, but $\mathsf{M}, s \not\models_{\mathrm{ELCDF}} C_G\psi$, then there are $s_i \in \mathsf{W}$, $a_i \in G$, $0 \le i \le n$ for some $n \in \mathbb{N}$ such that: $s_0 = s$, $\mathsf{M}, s_n \not\models \psi$ and $\mathsf{C}(a_i) \subseteq \mathsf{E}(s_{i-1}, s_i)$ for $1 \le i \le n$. Since $\mathsf{C}(a_i) \subseteq \mathsf{E}(s_{i-1}, s_i)$, we have either $\{\chi \mid D_H\chi \in tail(s_{i-1})\} \subseteq tail(s_i)$ for some H containing a_i or $\{\chi \mid F_{\{a_i\}}\chi \in tail(s_{i-1})\} \subseteq tail(s_i)$. In both cases, $\{\chi \mid K_{a_i}\chi \in tail(s_{i-1})\} \subseteq tail(s_i)$. Since $C_G\psi \in tail(s_i)$ implies $K_{a_i}C_G\psi, K_{a_i}\psi \in tail(s_i)$, we can infer that $C_G\psi, \psi \in tail(s_n)$. By the induction hypothesis, $\mathsf{M}, s_n \models_{\mathrm{ELCDF}} \psi$, leading to a contradiction.

Suppose $C_G\psi \notin tail(s)$, but $\mathsf{M}, s \models_{\mathrm{ELCDF}} C_G\psi$. Thus for any $s_i \in \mathsf{W}$ and $a_i \in G$ where $0 \le i \le n$, such that: $s_0 = s$ and $\mathsf{C}(a_i) \subseteq \mathsf{E}(s_{i-1}, s_i)$, we have $M, s_n \models_{\mathrm{ELCDF}} \psi$ and $M, s_n \models_{\mathrm{ELCDF}} C_G\psi$. Collect all such possible s_n above and s into the set $\mathcal{S}$; similarly collect all the $tail(s_n)$ and $tail(s)$ into the set Θ. We define $\delta = \bigvee_{t \in \mathcal{S}} \widehat{tail(t)}$, where for any $t \in \mathsf{W}$, $\widehat{tail(t)} = \bigwedge tail(t)$. (In general, for any finite set Ψ of formulas, we write $\widehat{\Psi}$ for $\bigwedge \Psi$.) We claim that $\vdash_{\mathrm{ELCDF}} \delta \to K_a\delta$ and $\vdash_{\mathrm{ELCDF}} \delta \to K_a\psi$ for any $a \in G$. By this claim and (C2) we have $\vdash_{\mathrm{ELCDF}} \delta \to C_G\psi$, and then by $\widehat{tail(s)} \to \delta$ we have $\widehat{tail(s)} \to C_G\psi$. In this way $C_G\psi \in tail(s)$, which leads to a contradiction. As for the proof of the claim:

(1) Suppose $\nvdash_{\mathrm{ELCDF}} \delta \to K_a\delta$, then $\delta \wedge \neg K_a\delta$ is consistent. Then there exists $t_0 \in \mathcal{S}$ such that $\widehat{tail(t_0)} \wedge \neg K_a\delta$ is consistent. Notice that $\vdash_{\mathrm{ELCDF}} \bigvee_{t \in \mathsf{W}} \widehat{tail(t)}$, hence we have a consistent set $\widehat{tail(t_0)} \wedge \neg K_a\neg\widehat{tail(t_1)}$ for some $t_1 \in \mathsf{W} \setminus \mathcal{S}$; for otherwise we have $\mathsf{W} \setminus \mathcal{S} = \emptyset$, hence $\vdash_{\mathrm{ELCDF}} \delta$, which leads to $\vdash_{\mathrm{ELCDF}} K_a\delta$, contradicting with $\nvdash_{\mathrm{ELCDF}} \delta \to K_a\delta$. Thus we have $\{\chi \mid K_a\chi \in tail(t_0)\} \subseteq tail(t_1)$, which implies $\{\chi \mid D_{\{a\}}\chi \in tail(t_0)\} \subseteq tail(t_1)$. Moreover, for any χ, if $D_{\{a\}}\chi \in tail(t_1)$, then $\widehat{tail(t_0)} \wedge \neg K_a\neg D_{\{a\}}\chi$ is consistent, hence $\widehat{tail(t_0)} \wedge \chi$ is also consistent, thus $\chi \in tail(t_0)$. So we have $\{\chi \mid D_{\{a\}}\chi \in tail(t_1)\} \subseteq tail(t_0)$. Now we let t_2 be t_0 extended with $\langle(\{a\}, d), tail(t_1)\rangle$, we have $\mathsf{C}(a) \subseteq \mathsf{E}(t_0, t_2)$. Hence $t_2 \in \mathcal{S}$ but $tail(t_2) = tail(t_1) \notin \Theta$. A contradiction!

(2) Suppose $\nvdash_{\mathrm{ELCDF}} \delta \to K_a\psi$, then $\delta \wedge \neg K_a\psi$ is consistent. So there exists $t_0 \in \mathcal{S}$ such that $\widehat{tail(t_0)} \wedge \neg K_a\psi$ is consistent. Thus $\{\sim\psi\} \cup \{\chi \mid D_{\{a\}}\chi \in tail(t_0)\} \cup \{\neg D_{\{a\}}\sim\chi \in cl(\theta) \mid \chi \in tail(t_0)\}$ is consistent. Hence it can be extended to some max consistent subset Δ^+ in $cl(\theta)$. Let t_1 be t_0 extended with $\langle(\{a\}, d), \Delta^+\rangle$, we have $\{\chi \mid D_{\{a\}}\chi \in tail(t_0)\} \subseteq tail(t_1)$. Moreover, if $D_{\{a\}}\chi \in tail(t_1) = \Delta^+$, then $\neg D_{\{a\}}\chi \notin \Delta^+$, thus $\neg D_{\{a\}}\sim\neg\chi = \neg D_{\{a\}}\chi \in cl(\theta)$ and $\neg\chi \notin tail(t_0)$, which implies $\chi \in tail(t_0)$. So we also have $\{\chi \mid D_{\{a\}}\chi \in tail(t_1)\} \subseteq tail(t_0)$. Thus we have $\mathsf{C}(a) \subseteq \mathsf{E}(t_0, t_1)$. Hence $t_1 \in \mathcal{S}$ and then $\mathsf{M}, t_1 \models_{\mathrm{ELCDF}} \psi$, which contradicts with $\sim\psi \in tail(t_1)$ by the

24

induction hypothesis. □

Theorem 3.17 ELCDF *is weakly complete for* ELCDF. □

Building on the discussion earlier in this section, the completeness proofs for **ELC**, **ELCD** and **ELCF** can be derived from the proofs for **ELCDF**. For brevity, however, the intricate details of these adaptations are not included in this paper.

4   Conclusion

We examined epistemic logics with all combinations of common, distributed and field knowledge, interpreted in scenarios that consider agents' epistemic abilities, such as professions. We adopted a type of similarity model that extends from a Kripke model by adding weights to edges, and studied the axiomatization of the resulting logics.

The framework of our logics presents diverse possibilities for characterizing the concept of *knowability*. Apart from interpreting knowability as known after a single announcement [4], a group announcement [2], or after a group resolves their knowledge [3], it is now conceivable to perceive knowability as known after an agent acquires certain skills (epistemic abilities) from some source or from a given group. Our framework also enables us to easily characterize *forgetability* or *degeneration* through changes in epistemic abilities, a process that is not as straightforward in classical epistemic logic.

Looking ahead, we aim to explore more sophisticated conditions on the similarity relation, such as those introduced in [6]. It would also be of interest to compare our framework with existing ones that use the same style of models, as presented in [17,7].

A   Proofs regarding Expressivity

In Figure 3, every language, with the exception of $\mathcal{ELCDF}$, has an arrow pointing to its immediate superlanguages. This is clearly true, as by definition, every language is at most as expressive as its superlanguages. In the case when $|\mathsf{Ag}| = 1$, a reverse arrow also exists between languages that either both contain common knowledge or neither contain common knowledge.

Lemma A.1 *When $|\mathsf{Ag}| = 1$ (i.e., when there is only one agent available in the language),*

(i) $\mathcal{EL} \equiv \mathcal{ELD} \equiv \mathcal{ELF} \equiv \mathcal{ELDF}$

(ii) $\mathcal{ELC} \equiv \mathcal{ELCD} \equiv \mathcal{ELCF} \equiv \mathcal{ELCDF}$

(iii) $\mathcal{ELDF} \prec \mathcal{ELC}$, *and hence any language in the first clause are less expressive than any language in the second clause.*

Proof. 1 & 2. In the case when $|\mathsf{Ag}| = 1$ there is only one agent, and since $D_{\{a\}}\varphi$ and $F_{\{a\}}\varphi$ are equivalent to $K_a\varphi$, the operators for distributed and mutual knowledge are redundant in this case. Hence the lemma.

3. We show that $\mathcal{ELC} \not\preceq \mathcal{ELDF}$, and so $\mathcal{ELDF} \prec \mathcal{ELC}$ since $\mathcal{ELDF} \equiv \mathcal{EL} \preceq \mathcal{ELC}$ by the first clause. Suppose towards a contradiction that there exists a

formula φ of $\mathcal{ELDF}$ equivalent to $C_{\{a\}}p$. Consider the set $\Phi = \{E^n_{\{a\}}p \mid n \in \mathbb{N}\} \cup \{\neg C_{\{a\}}p\}$. It is not hard to see that any finite subset of Φ is satisfiable, but not Φ itself. Let k be the length of φ (refer to a modal logic textbook for its definition), and suppose $\{E^n_{\{a\}}p \mid n \in \mathbb{N}, n \le k\} \cup \{\neg\varphi\}$ is satisfied at a state w in a model $M = (W, A, E, C, \nu)$. For any $s, t \in W$, we say that s reaches t in one step if $C(a) \subseteq E(s, t)$. Consider the model $M_k = (W_k, A, E, C, \nu)$, where W_k is set of states reachable from w in at most k steps. We can verify that $M_k, w \models \{E^n_{\{a\}}p \mid n \in \mathbb{N}\} \cup \{\neg\varphi\}$, which implies that Φ is satisfiable, leading to a contradiction. $\qquad\square$

We now proceed to elucidate the absence of arrows in the figure for the case when $|\mathsf{Ag}| \ge 2$.

Lemma A.2 *When $|\mathsf{Ag}| \ge 2$ (i.e., when there are at least two agents available in the language),*

(i) *For any superlanguage $\mathcal{L}$ of $\mathcal{ELC}$, and any sublanguage $\mathcal{L}'$ of $\mathcal{ELDF}$, it is not the case that $\mathcal{L} \preceq \mathcal{L}'$;*

(ii) *For any superlanguage $\mathcal{L}$ of $\mathcal{ELD}$, and any sublanguage $\mathcal{L}'$ of $\mathcal{ELCF}$, it is not the case that $\mathcal{L} \preceq \mathcal{L}'$;*

(iii) *For any superlanguage $\mathcal{L}$ of $\mathcal{ELF}$, and any sublanguage $\mathcal{L}'$ of $\mathcal{ELCD}$, it is not the case that $\mathcal{L} \preceq \mathcal{L}'$.*

Proof. 1. The Proof of Lemma A.1(iii) can be used here to show that $\mathcal{ELC} \not\preceq \mathcal{ELDF}$ also when $|\mathsf{Ag}| \ge 2$.

2. Consider models $M = (W, A, E, C, \nu)$ and $M' = (W', A, E', C, \nu')$, where $A = \{1, 2, 3\}$, $C(a) = \{1, 2\}$, $C(b) = \{1, 3\}$ (if there are more agents in the language, they are irrelevant here), and are illustrated below.

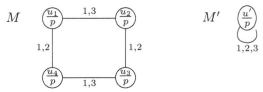

We can show by induction that for any formula φ of $\mathcal{ELCF}$, $M, u_1 \models \varphi$ iff $M', u' \models \varphi$. On the other hand, $M, u_1 \models D_{ab}\bot$ but $M', u' \not\models D_{ab}\bot$. It means that no $\mathcal{ELCF}$-formula can discern between M, u_1 and M', u', while languages with distributed knowledge can. Thus the lemma holds.

3. Consider similarity models $M = (W, A, E, C, \nu)$ and $M' = (W', A, E', C, \nu')$, where $A = \{1, 2, 3\}$, $C(a) = \{1, 2\}$, $C(b) = \{1, 3\}$, and are illustrated below.

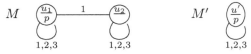

We can show by induction that for any formula φ of $\mathcal{ELCD}$, $M, u_1 \models \varphi$ iff $M', u' \models \varphi$. Meanwhile, we have $M, u_1 \not\models F_{ab}p$ and $M', u' \models F_{ab}p$. It follows that no $\mathcal{ELCD}$-formula can discern between M, u_1 and M', u', while languages with field knowledge can. Thus the lemma holds. □

As per Figure 3, Lemma A.2 suggests that there is not an arrow or a path of arrows leading from $\mathcal{ELC}$ (or any language having an arrow or a path of arrows originating from $\mathcal{ELC}$) to $\mathcal{ELDF}$ (or any language with an arrow or a path of arrows pointing to $\mathcal{ELDF}$). Similar relationships exist between $\mathcal{ELD}$ and $\mathcal{ELCF}$, and between $\mathcal{ELF}$ and $\mathcal{ELCD}$. Furthermore, in Figure 3, if there is an arrow or a path of arrows leading from one language to another, and not the other way round, this signifies that the first language is less expressive than the second. If there is no arrow or path of arrows in either direction between two languages, they are deemed incomparable. These observations lead us directly to the following corollary.

Corollary A.3 *When $|\mathsf{Ag}| \geq 2$,*

(i) $\mathcal{EL} \prec \mathcal{ELC}$, $\mathcal{ELD} \prec \mathcal{ELCD}$, $\mathcal{ELF} \prec \mathcal{ELCF}$ and $\mathcal{ELDF} \prec \mathcal{ELCDF}$;

(ii) $\mathcal{EL} \prec \mathcal{ELD}$, $\mathcal{ELC} \prec \mathcal{ELCD}$, $\mathcal{ELF} \prec \mathcal{ELDF}$ and $\mathcal{ELCF} \prec \mathcal{ELCDF}$;

(iii) $\mathcal{EL} \prec \mathcal{ELF}$, $\mathcal{ELC} \prec \mathcal{ELCF}$, $\mathcal{ELD} \prec \mathcal{ELDF}$ and $\mathcal{ELCD} \prec \mathcal{ELCDF}$;

(iv) $\mathcal{ELC}$, $\mathcal{ELD}$ and $\mathcal{ELF}$ are pairwise incomparable;

(v) $\mathcal{ELCD}$, $\mathcal{ELCF}$ and $\mathcal{ELDF}$ are pairwise incomparable;

(vi) $\mathcal{ELC}$ is incomparable with $\mathcal{ELDF}$;

(vii) $\mathcal{ELD}$ is incomparable with $\mathcal{ELCF}$;

(viii) $\mathcal{ELF}$ is incomparable with $\mathcal{ELCD}$.

B   Proof of Lemma 3.3

We provide a proof for Lemma 3.3 while first repeating it:

Lemma B.1 *The following hold:*

(i) *Given a symmetric Kripke model N, its translation N^ρ is a similarity model;*

(ii) *For any $\mathcal{ELCD}$-formula φ, any symmetric Kripke model N and any state s of N, $N, s \Vdash \varphi$ iff $N^\rho, s \models \varphi$.*

Proof. (i) Let $N = (W, R, V)$ be a symmetric Kripke model, and its translation $N^\rho = (W, \mathsf{Ag} \cup \{b\}, E, C, \nu)$. For any $a \in \mathsf{Ag} \cup \{b\}$ and $s, t \in W$, we have:

$$a \in E(s, t) \iff (s, t) \in R(a) \text{ (Def. 3.2)}$$
$$\iff (t, s) \in R(a) \text{ (since } R(a) \text{ is symmetric)}$$
$$\iff a \in E(t, s). \text{ (Def. 3.2)}$$

Hence N^ρ satisfies symmetry. Furthermore, N^ρ satisfies positivity, as there cannot be any $s, t \in W$ such that $E(s, t) = A \cup \{b\}$ and $s \neq t$. Hence N^ρ is a similarity model.

(ii) Let $N = (W, R, V)$ and its translation $N^\rho = (W, \mathsf{Ag} \cup \{b\}, E, C, \nu)$. We show the lemma by induction on φ. The cases involving atomic propositions, Boolean connectives, and common knowledge are straightforward to

27

verify since their semantic definitions follow a consistent pattern that facilitates smooth inductive reasoning. In this proof, we focus specifically on the cases for individual and distributed knowledge. It should be noted that the case for individual knowledge can be regarded as a particular instance of distributed knowledge; however, we include the details here for readers who seek a thorough clarity:

$$
\begin{aligned}
N, s \Vdash K_a \psi \iff & \text{ for all } t \in W, \text{ if } (s,t) \in R(a) \text{ then } N, t \Vdash \psi \\
\iff & \text{ for all } t \in W, \text{ if } a \in E(s,t) \text{ then } N, t \Vdash \psi \\
\iff & \text{ for all } t \in W, \text{ if } C(a) \subseteq E(s,t) \text{ then } N, t \Vdash \psi \\
\iff & \text{ for all } t \in W, \text{ if } C(a) \subseteq E(s,t) \text{ then } N^\rho, t \models \psi \\
\iff & N^\rho, s \models K_a \psi.
\end{aligned}
$$

$$
\begin{aligned}
N, s \Vdash D_G \psi \iff & \text{ for all } t \in W, \text{ if } (s,t) \in \bigcap_{a \in G} R(a), \text{ then } N, t \Vdash \psi \\
\iff & \text{ for all } t \in W, \text{ if } (s,t) \in R(a) \text{ for all } a \in G, \text{ then } N, t \Vdash \psi \\
\iff & \text{ for all } t \in W, \text{ if } C(a) \subseteq E(s,t) \text{ for all } a \in G, \text{ then } N, t \Vdash \psi \\
\iff & \text{ for all } t \in W, \text{ if } \bigcup_{a \in G} C(a) \subseteq E(s,t), \text{ then } N, t \models \psi \\
\iff & \text{ for all } t \in W, \text{ if } \bigcup_{a \in G} C(a) \subseteq E(s,t), \text{ then } N^\rho, t \models \psi \\
\iff & N^\rho, s \models D_G \psi.
\end{aligned}
$$

C   Completeness of EL by the Canonical Model Method

Lemma C.1 (canonicity) *The canonical model for* EL *is a similarity model.*

Proof. Let $M = (W, A, E, C, v)$ be the canonical model for EL. Notice that $\emptyset \notin C(a)$ for any agent a, so $E(s,t) \neq A$ for any $s, t \in W$, ensuring positivity. The symmetry of the model is evident as $E(s,t) = E(t,s)$ for any $s, t \in W$. Therefore, M is a similarity model. $\square$

Lemma C.2 (Truth Lemma) *Let* $M = (W, A, E, C, v)$ *be the canonical model for* EL. *For any* $\Gamma \in W$ *and any* $\mathcal{EL}$-*formula* φ, *we have* $\varphi \in \Gamma$ *iff* $M, \Gamma \models_{EL} \varphi$.

Proof. We will only demonstrate the case when φ is of the form $K_a \psi$ here.

Assuming $K_a \psi \in \Gamma$, but $M, \Gamma \not\models_{EL} K_a \psi$, there would exist a $\Delta \in W$ such that $C(a) \subseteq E(\Gamma, \Delta)$ and $M, \Delta \not\models_{EL} \psi$. Consequently, $\{\chi \mid K_a \chi \in \Gamma\} \subseteq \Delta$ (otherwise $\{a\} \notin E(\Gamma, \Delta)$, contradicting $\{a\} \in C(a)$). Thus, $\psi \in \Delta$. It follows from the induction hypothesis that $M, \Delta \models_{EL} \psi$, which results in a contradiction.

For the opposite direction, suppose $K_a \psi \notin \Gamma$, but $M, \Gamma \models_{EL} K_a \psi$, then for any $\Delta \in W$, $C(a) \subseteq E(\Gamma, \Delta)$ implies $M, \Delta \models_{EL} \psi$. First, we assert that $\{\neg \psi\} \cup \{\chi \mid K_a \chi \in \Gamma\} \cup \{\neg K_a \neg \chi \mid \chi \in \Gamma\}$ is **EL** consistent. If not, note that for any $\eta \in \{\chi \mid K_a \chi \in \Gamma\}$, we have $\neg K_a \neg K_a \eta \in \{\neg K_a \neg \chi \mid \chi \in \Gamma\}$. As $\vdash_{\textbf{EL}} \neg K_a \neg K_a \eta \to \eta$, it follows that $\{\neg \psi\} \cup \{\neg K_a \neg \chi \mid \chi \in \Gamma\}$ is not **EL** consistent. Therefore, we have $\vdash_{\textbf{EL}} (\bigwedge_{\chi \in \Gamma_0} \neg K_a \neg \chi) \to \psi$ for some finite subset Γ_0 of Γ. This leads to $\vdash_{\textbf{EL}} K_a ((\bigwedge_{\chi \in \Gamma_0} \neg K_a \neg \chi) \to \psi)$, and hence $\vdash_{\textbf{EL}} \bigwedge_{\chi \in \Gamma_0} K_a \neg K_a \neg \chi \to K_a \psi$. Since we have $\vdash_{\textbf{EL}} \chi \to K_a \neg K_a \neg \chi$ for any $\chi \in \Gamma_0$, it follows that we have $\vdash_{\textbf{EL}} (\bigwedge_{\chi \in \Gamma_0} \chi) \to K_a \psi$. This deduction implies that $K_a \psi \in \Gamma$, which contradicts our previous assumption. Now, let us extend the set $\{\neg\psi\} \cup \{\chi \mid K_a\chi \in \Gamma\} \cup \{\neg K_a\neg\chi \mid \chi \in \Gamma\}$ to some maximal **EL**-consistent set Δ^+ of $\mathcal{EL}$-formulas. Notice that $K_a\chi \in \Gamma$ implies $\chi \in \Delta^+$ for any χ. Furthermore, if we suppose $\chi \notin \Gamma$, then $\neg\chi \in \Gamma$, which leads to $\neg K_a\neg\neg\chi \in \Delta^+$, implying $\neg K_a\chi \in \Delta^+$. Therefore, $K_a\chi \in \Delta^+$ implies $\chi \in \Gamma$ for any χ. Given these stipulations, we find that $\mathsf{C}(a) \subseteq \mathsf{E}(\Gamma, \Delta)$. However, by using the induction hypothesis, we see that $M, \Delta \not\models_{\mathrm{EL}} \psi$. As a result, $M, \Gamma \not\models_{\mathrm{EL}} K_a\psi$. This conclusion contradicts our previous assumptions, confirming this direction of the lemma. □

With the Truth Lemma, we can state the following theorem:

Theorem C.3 (completeness of EL, with a direct proof) *For any $\mathcal{EL}$-formula φ and any set Φ of $\mathcal{EL}$-formulas, if $\Phi \models_{\mathrm{EL}} \varphi$, then $\Phi \vdash_{\mathbf{EL}} \varphi$.*

Proof. To prove this, suppose the contrary: $\Phi \not\vdash_{\mathbf{EL}} \varphi$. In this case, the set $\Phi \cup \{\neg\varphi\}$ can be extended to a maximal **EL**-consistent set Δ^+. In the canonical model for EL, denoted M, we have $M, \Delta^+ \models \chi$ for any formula $\chi \in \Phi \cup \{\neg\varphi\}$. This conclusion leads to $\Phi \not\models_{\mathrm{EL}} \varphi$. □

D   Truth Lemma for Theorem 3.14

Lemma D.1 *Let $M = (W, A, E, C, v)$ be the standard model for ELDF. For any $s \in W$ and any $\mathcal{ELDF}$-formula φ, $\varphi \in tail(s)$ if and only if $M, s \models_{\mathrm{ELDF}} \varphi$.*

Proof. The proof is by induction on φ, and we only display the cases for modalities.

Case $\varphi = K_a\psi$. Suppose $K_a\psi \in tail(s)$, but $M, s \not\models_{\mathrm{ELDF}} K_a\psi$, then there exists $t \in W$ such that $\mathsf{C}(a) \subseteq \mathsf{E}(s, t)$ and $M, t \not\models_{\mathrm{ELDF}} \psi$. Therefore, $\{\chi \mid D_G\chi \in tail(s)\} \subseteq tail(t)$ for some group G containing a or $\{\chi \mid X_{\{a\}}\chi \in tail(s)\} \subseteq tail(t)$. In both scenarios, $\psi \in tail(t)$ since $K_a\psi \in tail(s)$ implies $D_G\psi, F_{\{a\}}\psi \in tail(s)$. By the induction hypothesis, we have $M, t \models_{\mathrm{ELDF}} \psi$, which leads to a contradiction. Suppose $K_a\psi \notin tail(s)$, but $M, s \models_{\mathrm{ELDF}} K_a\psi$, then $\mathsf{C}(a) \subseteq \mathsf{E}(s, t)$ implies $M, t \models_{\mathrm{ELDF}} \psi$ for any $t \in W$. Extend $\{\neg\psi\} \cup \{\chi \mid K_a\chi \in tail(s)\} \cup \{\neg K_a\neg\chi \mid \chi \in tail(s)\}$ to some maximal **ELDF**-consistent set Δ^+, thus $\mathsf{C}(a) \subseteq \mathsf{E}(s, t)$ where t extends s with $\langle(\{a\}, d), \Delta^+\rangle$. By the induction hypothesis, we have $M, t \models_{\mathrm{ELDF}} \neg\psi$, which is contradictary.

Case $\varphi = D_G\psi$. Suppose $D_G\psi \in tail(s)$, but $M, s \not\models_{\mathrm{ELDF}} D_G\psi$, then there exists some $t \in W$ such that $\bigcup_{a \in G} \mathsf{C}(a) \subseteq \mathsf{E}(s, t)$ and $M, t \not\models_{\mathrm{ELDF}} \psi$. Therefore, $\{\chi \mid D_H\chi \in tail(s)\} \subseteq tail(t)$ for some group H such that $G \subseteq H$. We have $\psi \in tail(t)$ since $D_G\psi \in tail(s)$ implies $D_H\psi \in tail(s)$. By the induction hypothesis, we have $M, t \models_{\mathrm{ELDF}} \psi$, which leads to a contradiction. Suppose $D_G\psi \notin tail(s)$, but $M, s \models_{\mathrm{ELDF}} D_G\psi$, then $\bigcup_{a \in G} \mathsf{C}(a) \subseteq \mathsf{E}(s, t)$ implies $M, t \models_{\mathrm{ELDF}} \psi$ for any $t \in W$. Extend $\{\neg\psi\} \cup \{\chi \mid D_G\chi \in tail(s) \cup \{\neg D_G\neg\chi \mid \chi \in tail(s)\}\}$ to some maximal **ELDF**-consistent set Δ^+, thus $\bigcup_{a \in G} \mathsf{C}(a) \subseteq \mathsf{E}(s, t)$ where t extends s with $\langle(G, d), \Delta^+\rangle$. We have $M, t \models_{\mathrm{ELDF}} \neg\psi$ by the induction hypothesis. A contradiction!

Case $\varphi = F_G\psi$. Suppose $F_G\psi \in tail(s)$, but $M, s \not\models_{\mathrm{ELDF}} F_G\psi$, then there exists $t \in W$ such that $\bigcap_{a \in G} \mathsf{C}(a) \subseteq \mathsf{E}(s, t)$ and $M, t \not\models_{\mathrm{ELDF}} \psi$. Therefore,

$\{\chi \mid F_H\chi \in tail(s)\} \subseteq tail(t)$ for some group H such that $H \subseteq G$ or $\{\chi \mid D_J\chi \in tail(s)\} \subseteq tail(t)$ for some group J such that $G \cap J \neq \emptyset$. In both scenarios, we have $\psi \in tail(t)$ since $F_G\psi \in tail(s)$ implies $F_H\psi, D_J\psi \in tail(s)$. By the induction hypothesis, we have $\mathsf{M}, t \models_{\mathrm{ELDF}} \psi$, which leads to a contradiction. Suppose $F_G\psi \notin tail(s)$, but $\mathsf{M}, s \models_{\mathrm{ELDF}} F_G\psi$, then $\bigcap_{a \in G} \mathsf{C}(a) \subseteq \mathsf{E}(s, t)$ implies $\mathsf{M}, t \models_{\mathrm{ELDF}} \psi$ for any $t \in \mathsf{W}$. Extend $\{\neg\psi\} \cup \{\chi \mid F_G\chi \in tail(s)\} \cup \{\neg F_G\neg\chi \mid \chi \in tail(s)\}$ to some maximal **ELDF**-consistent set Δ^+, thus $\bigcap_{a \in G} \mathsf{C}(a) \subseteq \mathsf{E}(s, t)$ where t extends s with $\langle (G, m), \Delta^+ \rangle$. By the induction hypothesis, $\mathsf{M}, t \models_{\mathrm{ELDF}} \neg\psi$, which leads to a contradiction. $\qquad \square$

E   Extra Cases for the Proof of Lemma 3.16

The proof of Lemma 3.16 in the main text (p. 16) only contains the case for common knowledge. Here we supplement the cases for distributed and field knowledge for the careful reader (individual knowledge can be treated as a special case of the two).

Case $\varphi = D_G\psi$. The direction from $D_G\psi \in tail(s)$ to $\mathsf{M}, s \models_{\mathrm{ELCDF}} D_G\psi$ is similarly to the proof of Lemma D.1. For the other direction, suppose $D_G\psi \notin tail(s)$, but $\mathsf{M}, s \models_{\mathrm{ELCDF}} D_G\psi$, then $\bigcup_{a \in G} \mathsf{C}(a) \subseteq \mathsf{E}(s, t)$ implies $\mathsf{M}, t \models_{\mathrm{ELCDF}} \psi$ for any $t \in \mathsf{W}$. Notice that $\{\sim\psi\} \cup \{\chi \mid D_G\chi \in tail(t_0)\} \cup \{\neg D_G\sim\chi \in cl(\theta) \mid \chi \in tail(t_0)\}$ is a consistent subset of $cl(\theta)$. Extend it to a maximal **ELCDF**-consistent set Δ^+ in $cl(\theta)$. Thus, by a similar method to the proof of $\vdash_{\mathbf{ELCDF}} \delta \to K_a\psi$ in Lemma 3.16, we have $\{\chi \mid D_G\chi \in tail(s)\} \subseteq \Delta^+$ and $\{\chi \mid D_G\chi \in \Delta^+\} \subseteq tail(s)$. Let t be s extended with $\langle (G, d), \Delta^+ \rangle$, we have $\bigcup_{a \in G} \mathsf{C}(a) \subseteq \mathsf{E}(s, t)$. By the induction hypothesis we have $\mathsf{M}, t \not\models_{\mathrm{ELCDF}} \psi$, contradicting with $\mathsf{M}, s \models_{\mathrm{ELCDF}} D_G\psi$.

The case when $\varphi = F_G\psi$ is similar to the case for distributed knowledge except that we extend the consistent set $\{\sim\psi\} \cup \{\chi \mid F_G\chi \in tail(t_0)\} \cup \{\neg F_G\sim\chi \in cl(\theta) \mid \chi \in tail(t_0)\}$ to get a maximal Δ^+ in the closure, and let t be s extended with $\langle (G, m), \Delta^+ \rangle$.

References

[1] Aggarwal, C. C., "Data Mining: The Textbook," Springer, 2015.

[2] Ågotnes, T., P. Balbiani, H. van Ditmarsch and P. Seban, *Group announcement logic*, Journal of Applied Logic **8** (2010), pp. 62–81.

[3] Ågotnes, T. and Y. N. Wáng, *Resolving distributed knowledge*, Artificial Intelligence **252** (2017), pp. 1–21.

[4] Balbiani, P., A. Baltag, H. van Ditmarsch, A. Herzig, T. Hoshi and T. de Lima, *'Knowable' as 'known after an announcement'*, The Review of Symbolic Logic **1** (2008), pp. 305–334.

[5] Blackburn, P., M. de Rijke and Y. Venema, "Modal logic," Cambridge University Press, 2001.

[6] Chen, S., B. Ma and K. Zhang, *On the similarity metric and the distance metric*, Theoretical Computer Science **410** (2009), pp. 2365–2376.

[7] Dong, H., X. Li and Y. N. Wáng, *Weighted modal logic in epistemic and deontic contexts*, in: S. Ghosh and T. Icard, editors, *Proceedings of the Eighth International Conference on Logic, Rationality and Interaction (LORI 2021)*, Lecture Notes of Theoretical Computer Science **13039** (2021), pp. 73–87.

[8] Fagin, R., J. Y. Halpern, Y. Moses and M. Y. Vardi, "Reasoning about knowledge," The MIT Press, 1995.

[9] Fagin, R., J. Y. Halpern and M. Y. Vardi, *What can machines know? On the properties of knowledge in distributed systems*, Journal of the ACM **39** (1992), pp. 328–376.

[10] Halpern, J. Y. and Y. Moses, *A guide to completeness and complexity for modal logics of knowledge and belief*, Artificial Intelligence **54** (1992), pp. 319–379.

[11] Hansen, M., K. G. Larsen, R. Mardare and M. R. Pedersen, *Reasoning about bounds in weighted transition systems*, Logical Methods in Computer Science **14** (2018), pp. 1–32.

[12] Hintikka, J., "Knowledge and Belief: An Introduction to the Logic of Two Notions," Cornell University Press, Ithaca, New York, 1962.

[13] Larsen, K. G. and R. Mardare, *Complete proof systems for weighted modal logic*, Theoretical Computer Science **546** (2014), pp. 164–175.

[14] Larsen, K. G. and A. Skou, *Bisimulation through probabilistic testing*, Information and Computation **94** (1991), pp. 1–28.

[15] Liang, X. and Y. N. Wáng, *Epistemic logic via distance and similarity*, in: *Proceedings of PRICAI 2022: Trends in Artificial Intelligence* (2022), pp. 32–45.

[16] Meyer, J.-J. C. and W. van der Hoek, "Epistemic Logic for AI and Computer Science," Cambridge University Press, 1995.

[17] Naumov, P. and J. Tao, *Logic of confidence*, Synthese **192** (2015), pp. 1821–1838.

[18] Roelofsen, F., *Distributed knowledge*, Journal of Applied Non-Classical Logics **16** (2007), pp. 255–273.

[19] Sahlqvist, H., *Completeness and correspondence in the first and second order semantics for modal logic*, in: S. Kanger, editor, *Proceedings of the Third Scandinavian Logic Symposium*, Studies in Logic and the Foundations of Mathematics **82**, Elsevier, 1975 pp. 110 – 143.

[20] Tan, P.-N., M. Steinbach and V. Kumar, "Introduction to data mining," Pearson, 2005.

[21] van Ditmarsch, H., W. van der Hoek and B. Kooi, "Dynamic Epistemic Logic," Synthese Library **337**, Springer Netherlands, 2008.

[22] Wáng, Y. N. and T. Ågotnes, *Simpler completeness proofs for modal logics with intersection*, in: M. A. Martins and I. Sedlár, editors, *Dynamic Logic: New Trends and Applications*, Lecture Notes in Computer Science **12569** (2020), pp. 259–276.

Neuro-Symbolic Approach for Legal Tasks Based on Large Language Models

Bin Wei

Guanghua law school, Zhejiang University

Law&AI Lab, Zhejiang University

Abstract

In this paper, we introduce a neuro-symbolic approach for legal tasks, combining the strengths of large language models (LLMs) and classical symbolic methods to enhance the efficiency, accuracy, and interpretability of pre-litigation mediation and legal outcome prediction. This approach addresses the challenges posed by overwhelming caseloads and the complexity of legal texts.

The first component of the framework focuses on pre-litigation mediation. A hybrid model integrating LLMs, such as GPT-4, and classical machine learning algorithms is employed. The process begins with the vectorization of case data using OpenAI's embedding models to construct a dynamic legal knowledge base, facilitating the distinction between simple and hard cases through similarity comparisons. For undetermined cases, GPT-4 is used to extract key elements, followed by outcome prediction using traditional machine learning models. Bert-type models are then applied for the final classification, ensuring high accuracy and recall rates.

The second component extends to legal outcome prediction. Deep neural networks and human-in-the-loop inputs are used for accurate classification of petitions. This involves the use of supervised learning methods to handle the diverse and complex nature of legal documents. The integration of LLMs enhances the extraction and analysis of semantic information from extensive legal texts, improving the system's robustness and adaptability.

Experimental results validate the effectiveness of the hybrid framework, demonstrating superior performance in terms of accuracy, precision, and interpretability. This approach not only improves the decision-making process in pre-litigation mediation but also offers a promising solution for broader legal tasks. The neuro-symbolic framework represents a significant advancement in the application of AI to law, combining the strengths of LLMs and symbolic reasoning to address practical challenges in legal adjudication.

Keywords: neuro-symbolic approach, large language models, pre-litigation mediation, legal outcome prediction, GPT-4, Bert-type models, interpretability

A Principle-based Analysis for Numerical Balancing

Aleks Knoks[a,b] Muyun Shao[c] Leendert van der Torre[b,c]
Vincent de Wit[b] Liuwen Yu[b]

[a] *University of Luxembourg*
Institute of Philosophy
Esch-sur-Alzette, Luxembourg

[b] *University of Luxembourg*
Department of Computer Science
Esch-sur-Alzette, Luxembourg

[c] *Zhejiang University*
School of Philosophy
Hangzhou, China

Abstract

The more recent philosophical literature concerned with foundational questions about normativity often appeals to the notion of normative reasons, or considerations that count in favor or against actions, and their interaction. The interaction between reasons is standardly conceived of in terms of weighing reasons on (normative) weight scales. Knoks and van der Torre [8] have recently proposed a formal framework that allows one to think about the interaction between reasons as a kind of inference pattern. This paper extends that framework by introducing and exploring what we call *numerical balancing operators*. These operators represent the weights or magnitudes of reasons by means of numbers, and they are particularly well-suited for capturing the intuition of aggregating and weighing reasons. We define a number of concrete classes of balancing operators and explore them using a principle-based analysis.

Keywords: reasons, weighing, detachment, principle-based analysis.

1 Introduction

The notion of *normative reasons* has been playing an increasingly important role in the philosophical literature tackling foundational questions about normativity. In the practical domain, normative reasons are standardly understood to be facts that speak in favor of or against actions. [1] Thus, the fact that you have made a promise to a friend is a reason that speaks in favor of your keeping the promise, and the fact that throwing this punch would result in harming

---

[1] The locus classicus is Scanlon [13, p. 17]. See also [11], [12], [19], among many others.

someone is a reason that speaks against throwing the punch. The interaction between normative reasons is standardly taken to determine the deontic statuses of actions—whether they are permissible, obligatory, or forbidden—and this interaction itself is usually made sense of by analogy with weight scales. [2] On the simplest construction, these weight scales work roughly as follows. The reasons that speak in favor of some action φ (positive reasons) go in one pan of the scales, while those that speak against φ (negative reasons) go in the other. If the overall weight (or magnitude) of reasons in the first pan is greater than the overall weight of reasons in the second, φ is obligatory. If the overall weight of reasons in the second pan is greater, φ is forbidden. If the pans are equally balanced, φ is optional, that is, both φ and not-φ are permissible. [3]

While most of the work theorizing about the interaction between normative reasons and their relation to the overall deontic statuses of actions has been carried out informally, there are some exceptions. One such is a recent paper of Knoks and van der Torre [8]. [4] Our main goal in this paper is to adjust the approach of Knoks and van der Torre and apply it to (richer) structures in which reasons are associated with numerical weights and deontic statuses are assigned to actions on the basis of these weights—with this, the approach is steered closer to the way the interaction between reasons is conceived of in the informal (philosophical) literature. To reach our goal, we introduce the formal notion of *numerical balancing operators*, formulate some concrete classes of such operators, and carry out a principle-based analysis of them. The results we present in this paper show that adding numerical weights to the picture makes a huge difference: some of the core principles formulated in [8] no longer hold in general, and new principles need to be formulated to distinguish the operators.

The rest of this paper is structured as follows. In Section 2, we recall some basic notions from Knoks and van der Torre [8]. In Section 3, we extend the framework with numerical weights and introduce the core notion of numerical balancing operators. In Section 4, we introduce six concrete classes of balancing operators, and in Section 5, we present our principle-based analysis. Section 6 clarifies the relationship between the results we present here and the more general framework of Knoks and van der Torre [8]. Finally, Section 7 concludes and hints at some ideas for future research.

2 Preliminaries

In this section, we recall some definitions from [8]. The two basic building blocks in that paper are an infinite set $\mathcal{A}$ and an abstract set of values $\mathcal{V}$. Given that we will be interested in what Knoks and van der Torre call *balancing operations*,

---

[2] See, for instance, [1], [2], [9], [15], [17], [18].

[3] For the most careful (informal) analysis of the weight scales metaphor, see [18], for a good introduction, see [9].

[4] Other notable exceptions include Horty's [5], [6] default logic-based framework and the recent approaches that draw on decision and probably theory [3], [10], [14].

we will work with a concrete set of values, namely, $\{+, -, 0\}$. Our formal notion of a *reason* is then defined thus:

Definition 2.1 [Reasons] Let $\mathcal{A}$ be an infinite set, called the *universe of discourse* and let $\mathcal{V}$ be the set $\{+, -, 0\}$, called *values*. A reason r is a triple of the from (x, y, v) where x and y are elements of $\mathcal{A}$ and $v \in \{+, -\}$ is the value associated with a reason, also called the *polarity* of r. [5]

The next important notion is that of a *context*:

Definition 2.2 [Contexts] A context c is a pair of the form (R, y) where R is a finite set of reasons, and y is an element of $\mathcal{A}$, called the *issue*.

Contexts are meant to represent particular scenarios or situations. Each context can be thought of as asking a question about some action—this is why we call y an *issue*: is it the case that y ought to be taken, that y ought not to be taken, or that it is permissible to take y and also not to take it (that is, y is optional)? The set of reasons R of a context, in turn, is comprised of the considerations that are relevant for answering this question. We use $\mathcal{U}$ to denote the set of all possible contexts, that is, the set of contexts that can be constructed by Definitions 2.1–2.2.

Formally, balancing operations are functional relations between contexts and values. Intuitively, they can be thought of as answers to questions posed by contexts. If the context (R, y) is assigned a $+$, then y ought to be taken. If it is assigned a $-$, then y ought not to be taken. And if it is assigned a 0, then y is optional.

Definition 2.3 [Balancing operations] A *balancing operation*, denoted by $\mathcal{B}$, is a functional relation between contexts and values, that is, $\mathcal{B} \subseteq \mathcal{U} \times \mathcal{V}$ such that, for any $(c, v), (c', v') \in \mathcal{B}$, if $c = c'$, then $v = v'$.

With Definition 2.3 on the table, we are in a position to formulate principles that balancing operations might satisfy. Before we recall some important principles from [8], however, let us introduce some useful notation:

- Where $v \in \{+, 0, -\}$, we let $\overline{v}$ stand for the value that is opposite to v, that is: $\overline{v} = -$ if $v = +$; $\overline{v} = +$ if $v = -$; and $\overline{v} = 0$ if $v = 0$.

- Where $r = (x, y, v)$ is a reason, let $action(r) = y$ and $polarity(r) = v$.

- Where R is a set of reasons and $y \in \mathcal{A}$, the set of reasons from R that *speak in favor of* y is the set $pos(R, y) = \{r \in R : r = (x, y, +)\}$; the set of reasons from R that *speak against* y is the set $neg(R, y) = \{r \in R : r = (x, y, -)\}$; and the set of reasons from R that are relevant to y is the set $R_y = pos(R, y) \cup neg(R, y)$.

- When talking about sets of contexts, we can distinguish between the set of all possible contexts, denoted by $\mathcal{U}$, and the set of contexts under consideration,

---

[5] The reader familiar with the philosophical literature on reasons may notice that our technical concept corresponds to what is often called the *reason relation*.

denoted by $\mathcal{C}$. The latter is the set of contexts for which the balancing operation that we are discussing at a given point is defined.

While Knoks and van der Torre formulate a handful of principles, here we recall the two that, they claim, are particularly basic, intuitive, and important because they formalize properties that seem to be inherent in the metaphor of weighing reasons on scales. These principles are Relevance and Neutrality. The intuitive idea behind Relevance is that the values assigned to an issue y within a context must be based only on the reasons that are directly related to y, and, thus, that reasons that are not related to y can be removed from the context without affecting the result.

Principle 2.4 (Relevance) *A balancing operation $\mathcal{B}$ satisfies* Relevance *just in case if $((R, y), v) \in \mathcal{B}$ and $((R_y, y), v') \in \mathcal{B}$, then $v = v'$.*[6]

Turning to Neutrality, it is meant to capture the intuition that the values $+$ and $-$ should be treated equally: if we switch the polarities of all reasons in a given context, then the value that is assigned to the context should also switch.

Principle 2.5 (Neutrality) *Given a set of reasons R, let $R' = \{(x, y, \overline{v}) : (x, y, v) \in R\}$. A balancing operation $\mathcal{B}$ satisfies* Neutrality *just in case if $((R, y), v) \in \mathcal{B}$ and $((R', y), v') \in \mathcal{B}$, then $v' = \overline{v}$.*

3 Numerical balancing operators

An important part of the intuitive picture of weighing reasons on scales is that one reason can have more weight than another, and that the weights of multiple reasons can add up. The formal notion of balancing operations does not allow us to represent this idea explicitly. The main goal of this section then is to formulate an analogous notion—that of (numerical) *balancing operators*—that will allow us to do that.

As a first step, we introduce the notion of a *weight system*.

Definition 3.1 [Weight systems] Let $\mathcal{C}$ be a set of contexts. The set of reason-context pairs of $\mathcal{C}$, written as $\mathcal{X}_\mathcal{C}$, is the set $\{(r, (R, y)) : r \in R, (R, y) \in \mathcal{C}\}$. Then a *weight system for $\mathcal{C}$*, written as $w_\mathcal{C}$, is a pair (W, f_w) where $W \subseteq \mathbb{R}^+$ is a set of weights and $f_w : \mathcal{X}_\mathcal{C} \to W$ is a total function.

It is natural to wonder about the effects of context shifts on the weights of reasons, or to ask whether any given reason has to have the same weight in every context. The positions that have been explored in the philosophical literature here range from extreme *atomist views*, on which any given reason's

---

[6] Notice that, in general, a balancing operation $\mathcal{B}$ can be such that $((R, y), v) \in \mathcal{B}$, while $((R_y, y), v') \notin \mathcal{B}$. It's not difficult to define a constraint that rules out this possibility. We can think of it as a variation on Relevance. **Principle (Relevance')**: A balancing operation $\mathcal{B}$ satisfies *Relevance'* just in case if $((R, y), v) \in \mathcal{B}$, then there exists some value v' such that $((R_y, y), v') \in \mathcal{B}$. It shouldn't be difficult to see that Relevance and Relevance' entail the following stronger principle. **Principle (Relevance'')**: A balancing operation $\mathcal{B}$ satisfies *Relevance''*, Re'', just in case if $((R, y), v) \in \mathcal{B}$, then $((R_y, y), v) \in \mathcal{B}$.

weight and polarity are context-independent, to extreme *holist views*, on which a reason's weight and its polarity can both change from context to context. Since ours is a general and formal exploration, we do not want to commit to any particular view here. However, we also want to be able to express any view lying on the atomism-holism spectrum formally. While the above definition allows reasons to be associated with different weights in different contexts—naturally inviting a holist picture—we can impose further constraints on weight systems to express views that are closer to the atomist side of the spectrum. Thus, our next definition captures one of the core tenets of atomism: that the weights of reasons are context-independent.

Definition 3.2 [Fixed weight systems] Let C be a set of contexts and $w_C = (W, f_w)$ a weight system for C. Then w_C is called a *fixed weight system* just in case, for any reason r and any pair of contexts c, $c' \in C$, we have $f_w(r, c) = f_w(r, c')$.

The scales metaphor has it that the weights of individual reasons with the same polarity get aggregated into a collective weight, and that the collective weights of positive and negative reasons determine the final position of scales. Since in our framework this final position corresponds to the value associated with a context, we need a bridge from contexts supplemented with weight systems to values. This bridge is provided by what we call *procedures*:

Definition 3.3 [Procedures] Let $\mathcal{U}$ be the set of all contexts and $\mathcal{W}$ the of set of all weight systems for $\mathcal{U}$. A *procedure* is a function $\mathcal{P} : \mathcal{U} \times \mathcal{W} \to \mathcal{V}$ associating contexts and weight systems with values.

Notice that procedures are independent of weight systems: we can apply the same procedure to contexts with different weight systems, or different procedures to contexts with the same weight system.

Now we have all the ingredients we need to define balancing operators (our substitute for balancing operations from [8]). These, in effect, combine weight systems and procedures:

Definition 3.4 [Balancing operators] A *balancing operator*, denoted by $\mathcal{B}_o$, is a triple $(C, w_C, \mathcal{P})$ where C is a set of contexts, w_C a weight system for C, and $\mathcal{P}$ a procedure.

In the next section, we introduce several concrete (classes of) balancing operators. Before we turn to them, however, let's formulate some general principles that balancing operators can satisfy, and we start by restating Relevance and Neutrality from Section 2 as principles for balancing operators:

Principle 3.5 (Relevance) *A balancing operator* $\mathcal{B}_o = (C, w_C, \mathcal{P})$ *satisfies Relevance,* **Re***, just in case if there are v and v' such that* $\mathcal{P}((R, y), w_C) = v$ *and* $\mathcal{P}((R_y, y), w_C) = v'$, *then* $v = v'$.[7]

---

[7] The counterpart of the stronger version of Relevance discussed in footnote 6 would run as follows. **Principle (Relevance'')**: A balancing operator $\mathcal{B}_o = (C, w_C, \mathcal{P})$ satisfies *Relevance''*, **Re''**, just in case if $\mathcal{P}((R, y), w_C) = v$, then $\mathcal{P}((R_y, y), w_C) = v$.

Principle 3.6 (Neutrality) *Given a set of reasons R, let $R' = \{(x, y, \overline{v}) : (x, y, v) \in R\}$. A balancing operator $(\mathcal{C}, w_{\mathcal{C}}, \mathcal{P})$ satisfies* Neutrality, *Ne, just in case if $\mathcal{P}((R, y), w_{\mathcal{C}}) = v$ and $\mathcal{P}((R', y), w_{\mathcal{C}}) = v'$, then $v' = \overline{v}$.*

Recall our definition of fixed weight systems. We can use it to formulate another principle or constraint on balancing operators:

Principle 3.7 (Fixed Weight) *A balancing operator $\mathcal{B}_o = (\mathcal{C}, w_{\mathcal{C}}, \mathcal{P})$ satisfies* Fixed Weight, *FiWe, just in case $w_{\mathcal{C}}$ is a fixed weight system.*

According to atomism, not only the weights of reasons are fixed, but also their polarity. This idea can, again, be expressed in the form of a principle:

Principle 3.8 (Fixed Polarity) *A balancing operator $\mathcal{B}_o = (\mathcal{C}, w_{\mathcal{C}}, \mathcal{P})$ satisfies* Fixed Polarity, *FiPo, just in case, for any reason $r = (x, y, v)$, if there is a context $(R, y) \in \mathcal{C}$ such that $r \in R$, then there is no $(R', y) \in \mathcal{C}$ such that $(x, y, \overline{v}) \in R'$.*

With these two principles, we can formulate extreme atomism as a class of balancing operators.

Definition 3.9 [Atomist balancing operators] Let $\mathcal{B}_o$ be a balancing operator. We call $\mathcal{B}_o$ *atomist* just in case $\mathcal{B}_o$ satisfies both Fixed Polarity and Fixed Weight.

And given that holism is defined in opposition to atomism, it is also straightforward to formulate.

Definition 3.10 [Holist balancing operators] Let $\mathcal{B}_o$ be a balancing operator. We call $\mathcal{B}_o$ *holist* just in case it is not *atomist*.

It's worth noting that Fixed Weight and Fixed Polarity illustrate the flexibility of the formal notion of a balancing operator: we can formulate different principles some of which have to do with weight systems, others with the structure of contexts, and yet others with procedures.

4   Some concrete balancing operators

In this section, we introduce six classes of balancing operators. The unifying element of each class is the procedure. The first three classes correspond to three simple and intuitive operations on numbers: addition, multiplication and maximum. The forth class supplements the first of these with a threshold. Finally, the ideas behind our last two operators come from the discussion of (possible) views one might have about the workings of weight scales in Tucker [16].

The first class of operators is based on simple addition. The context (R, y) gets assigned the value $+$ if the sum weight of reasons for y is strictly greater than the sum weight of reasons against y; it gets assigned $-$ if the sum weight of reasons against y is strictly greater than the sum weight of reasons for y; and it gets assigned 0 otherwise.

Definition 4.1 [Additive Balancing Operators] Let $\mathcal{B}_o = (\mathcal{C}, w_{\mathcal{C}}, \mathcal{P}^+)$ be a balancing operator. Then it is called an *Additive Balancing Operator*, Add, just in case:

$$\mathcal{P}^+((R,y), w_{\mathcal{C}}) = \begin{cases} + & \text{if } \sum_{r \in pos(R,y)} f_w(r,(R,y)) > \sum_{r \in neg(R,y)} f_w(r,(R,y)) \\ - & \text{if } \sum_{r \in pos(R,y)} f_w(r,(R,y)) < \sum_{r \in neg(R,y)} f_w(r,(R,y)) \\ 0 & \text{otherwise} \end{cases}$$

The second class of balancing operators is based on multiplication. A context gets assigned $+$ in case the product of weights of positive reasons (for y) is greater than that of negative reasons; it gets assigned $-$ in case the product of weights of negative reasons is greater than that of positive reasons; and it gets assigned 0 otherwise.

Definition 4.2 [Multiplicative Balancing Operators] Let $\mathcal{B}_o = (\mathcal{C}, w_{\mathcal{C}}, \mathcal{P}^\times)$ be a balancing operator. Then it is called a *Multiplicative Balancing Operator*, Mul, just in case :

$$\mathcal{P}^\times((R,y), w_{\mathcal{C}}) = \begin{cases} + & \text{if } \prod_{r \in pos(R,y)} f_w(r,(R,y)) > \prod_{r \in neg(R,y)} f_w(r,(R,y)) \\ - & \text{if } \prod_{r \in pos(R,y)} f_w(r,(R,y)) < \prod_{r \in neg(R,y)} f_w(r,(R,y)) \\ 0 & \text{otherwise} \end{cases}$$

The balancing operators belonging to the third class we discuss determine the value of context by comparing the maximal weights of positive and negative reasons. A context gets assigned $+$ if the maximal weight of positive reasons is greater than that of negative reasons; it gets assigned $-$ if the maximal weight of negative reasons is greater than that of positive reasons; and it gets assigned 0 if the weights are equal.

Definition 4.3 [Maximizing Balancing Operators] Let $\mathcal{B}_o = (\mathcal{C}, w_{\mathcal{C}}, \mathcal{P}^m)$ be a balancing operator. Then it is called a *Maximizing Balancing Operator*, Max, just in case:

$$\mathcal{P}^m((R,y), w_{\mathcal{C}}) = \begin{cases} + & \text{if } \mathbf{Max}(\{f_w(r,(R,y)) : r \in pos(R,y)\}) > \\ & \qquad \mathbf{Max}(\{f_w(r,(R,y)) : r \in neg(R,y)\}) \\ - & \text{if } \mathbf{Max}(\{f_w(r,(R,y)) : r \in pos(R,y)\}) < \\ & \qquad \mathbf{Max}(\{f_w(r,(R,y)) : r \in neg(R,y)\}) \\ 0 & \text{otherwise} \end{cases}$$

The balancing operators that belong to the fourth class work with a *threshold* on the weights of reasons. The basic idea here is that a reason can make a difference for which value gets assigned to a context only in case its weight is above a certain threshold. In the following definition, this idea is combined with the familiar operation of addition:

Definition 4.4 [(Additive) Threshold Balancing Operators] Let $\mathcal{B}_o = (\mathcal{C}, w_{\mathcal{C}}, \mathcal{P}^t)$ be a balancing operator. Then it is called an *(Additive) Threshold Balancing Operator*, (Add)Thr, just in case:

$$\mathcal{P}^t((R,y), w_{\mathcal{C}}) = \begin{cases} + & \text{if } \sum_{r \in pos(R,y) \wedge f_w(r,(R,y)) > t} f_w(r,(R,y)) > \\ & \qquad \sum_{r \in neg(R,y) \wedge f_w(r,(R,y)) > t} f_w(r,(R,y)) \\ - & \text{if } \sum_{r \in pos(R,y) \wedge f_w(r,(R,y)) > t} f_w(r,(R,y)) < \\ & \qquad \sum_{r \in neg(R,y) \wedge f_w(r,(R,y)) > t} f_w(r,(R,y)) \\ 0 & \text{otherwise} \end{cases}$$

It's worth emphasizing that a threshold is not an operation, but, rather, a gatekeeping device that precludes reasons with (relatively) low weights from having any effect on the value assigned to a context. Definition 4.4 adds a threshold to addition. It should be clear that the operations of multiplication and taking the maximum that we used to define balancing operators above can also be supplemented with a threshold.

Now we turn to the final two classes of balancing operators. Both of these are inspired by the discussion in Tucker [16], who works in an informal setting and formulates the counterparts of our balancing operators in terms of *permission*. Since we have been working with obligations above, we re-state Tucker's ideas in terms of obligations.

The first of these two classes formalizes what Tucker calls *relative weight satisficing*: φ is permissible just in case the reasons against φ are no more than twice as weighty as the reasons for φ.[8] Restating this idea in terms of obligations, we get the following: φ is obligatory just in case the reasons against φ are *at most* twice as weighty as the reasons for φ *and* the reasons for φ are (strictly) *more than* twice as weighty as the reasons against φ. Since the second conjunct entails the first, we can simplify: φ is obligatory just in case the reasons for φ are *more than* twice as weighty as the reasons against φ. The formal definition, then, runs as follows:

Definition 4.5 [Relative Weight Satisficing Operators] Let $\mathcal{B}_o = (\mathcal{C}, w_{\mathcal{C}}, \mathcal{P}^R)$ be a balancing operator. Then it is called a *Relative Weight Satisficing Operator*, RelSat, just in case:

$$\mathcal{P}^R((R,y), w_{\mathcal{C}}) = \begin{cases} + & \text{if } \sum_{r \in pos(R,y)} f_w(r,(R,y)) > 2 \sum_{r \in neg(R,y)} f_w(r,(R,y)) \\ - & \text{if } 2 \sum_{r \in pos(R,y)} f_w(r,(R,y)) < \sum_{r \in neg(R,y)} f_w(r,(R,y)) \\ 0 & \text{otherwise} \end{cases}$$

Our final class of balancing operators corresponds to what Tucker calls *absolute weight satisficing*. (This view is meant to be in tension with the idea

---

[8] See [16, p. 373ff].

of weighing reasons on weight scales.) According to absolute weight satisficing, φ is permissible if the reasons for φ have a weight of at least 100 (no matter how much weight the reasons against φ have), and it is not permissible otherwise. [9] Notice that it is straightforward to define a similar sort of operator—that is, an operator that is sensitive to positive reasons *only*—in terms of obligations: φ is obligatory if the reasons for φ have a weight of at least 100, and it is forbidden (impermissible) otherwise. The formal definition then runs thus:

Definition 4.6 [Absolute Weight Satisficing Operators] Let $\mathcal{B}_o = (\mathcal{C}, w_\mathcal{C}, \mathcal{P}^A)$ be a balancing operator. Then it is called an *Absolute Weight Satisficing Operator*, AbsSat, just in case:

$$\mathcal{P}^A((R,y), w_\mathcal{C}) = \begin{cases} + & \text{if } \sum_{r \in pos(R,y)} f_w(r, (R,y)) > 100 \\ - & \text{otherwise} \end{cases}$$

Perhaps, one note about the final two class of operators is in order before we leave this section: we followed Tucker in setting the threshold at 100 in $\mathcal{P}^A$, as well as in requiring that the reasons against φ cannot be more than *two times* as weighty as the reasons for φ to be permissible in $\mathcal{P}^R$. We could define versions of these operators using other numbers.

5   Principle-based analysis

In this section, we formulate four principles and use them to compare the balancing operators defined in Section 4. We start with the formulation of the principles—the first comes from Knoks and van der Torre [8]; the latter three are new. Then we turn to a discussion, of our results and some complications.

The first principle is *Polarity Monotony*. It says that, if a balancing operator assigns $+$ to a context and a positive reason is added, then the operator will still assign $+$ to the context; and similarly, if the operator assigns $-$ to a context and a negative reason is added, then it will still assign $-$ to the context.

Principle 5.1 (Polarity Monotony) *A balancing operator* $\mathcal{B}_o = (\mathcal{C}, w_\mathcal{C}, \mathcal{P})$ *satisfies* Polarity Monotony, *PoMn, just in case, for all* $\mathcal{P}((R,y), w_\mathcal{C}) = v$ *where* $v \neq 0$, *if* $(R \cup \{(x, y, v)\}, y) \in \mathcal{C}$, *then* $\mathcal{P}((R \cup \{(x, y, v)\}, y), w_\mathcal{C}) = v$.

Our second principle is called *Commensurate Removal*. It says that for every context, if we remove a pair of opposite reasons with the same weight, then the value assigned to the context doesn't change.

Principle 5.2 (Commensurate Removal) *A balancing operator* $\mathcal{B}_o = (\mathcal{C}, w_\mathcal{C}, \mathcal{P})$ *satisfies* Commensurate Removal, *CoRe, just in case, if* $\mathcal{P}((R,y), w_\mathcal{C}) = v$, *then for each pair of reasons* $r, r' \in R$ *such that* $polarity(r) = polarity(r')$ *and* $f_w(r, (R,y)) = f_w(r', (R,y))$, *we have* $\mathcal{P}((R \backslash \{r, r'\}, y), w_\mathcal{C}) = v$.

Our next principle is called *Sensitivity*. It says that, for every equally-balanced context—that is, every context to which 0 is assigned—adding a new

---

[9] See [16, p. 378ff].

reason will cause the new context to be assigned a value that equals the polarity of that reason.

Principle 5.3 (Sensitivity) *A balancing operator* $\mathcal{B}_o = (\mathcal{C}, w_\mathcal{C}, \mathcal{P})$ *satisfies Sensitivity,* $\mathcal{S}e$, *just in case, if* $\mathcal{P}((R, y), w_\mathcal{C}) = 0$ *and there is a* v *such that* $\mathcal{P}((R \cup \{r\}, y), w_\mathcal{C}) = v$, *then* $v = polarity(r)$.

Finally, our final principle is *Union Monotony*. It says that if a balancing operator assigns the value v to two contexts, then it will also assign v to the union of these contexts.

Principle 5.4 (Union Monotony) *A balancing operator* $\mathcal{B}_o = (\mathcal{C}, w_\mathcal{C}, \mathcal{P})$ *satisfies* Union Monotony, *UnMn, just in case, if* $\mathcal{P}((R_1, y), w_\mathcal{C}) = v$, $\mathcal{P}((R_2, y), w_\mathcal{C}) = v$, *and* $(R_1 \cup R_2, y) \in \mathcal{C}$, *then* $\mathcal{P}((R_1 \cup R_2, y), w_\mathcal{C}) = v$.

Now that we have the principles, they can be used to analyze and compare the operators. However, there is a complication: the framework that we have set up is so unconstrained that, in the general case, (almost) none of the principles are satisfied by any of the operators. [10] This has to do, in particular, with the fact that our formal notion of a weight system (Definition 3.1) allows for unconstrained change of reasons' weights from one context to another. But let's recall our (brief) discussion of atomism and holism from Section 3 here. Atomists say that reasons weights and polarities are the same in all contexts, whereas extreme holists say that the weights of the same reason in two contexts can be wildly different. We wanted to be in a position to express all sorts of views lying on the atomism-holism spectrum in our framework, and, without imposing further constraints, it effectively imposes an extreme holist picture. On reflection, it should be no surprise that the balancing operators from Section 4 do not satisfy any of the principles *if* extreme holism is at work in the background.

What we present below then is a principle-based analysis of those balancing operators from Section 4 which also satisfy Fixed Weight, that is, we restrict attention to balancing operators with fixed weight systems.

Proving that a given operator does (or does not) satisfy some principle is more tedious than difficult. Here are two sample proofs:

Proposition 5.5 *Relative weight satisficing (Definition 4.5) with Fixed Weight (Principle 3.7) does not satisfy Sensitivity (Principle 5.3).*

Proof. Consider a relative weight satisficing operator $(\mathcal{C}, w_\mathcal{C}, \mathcal{P}^R)$ where $\mathcal{C} = \{c_1, c_2\}$; $c_1 = (\{r_1, r_2\}, y_1)$, $c_2 = (\{r_1, r_2, r_3\}, y_1)$; $r_1 = (x_1, y_1, +)$, $r_2 = (x_2, y_1, -)$, $r_3 = (x_3, y_1, +)$; $f_w(r_1, c_1) = f_w(r_2, c_1) = 5$, and $f_w(r_3, c_2) = 0.5$. Notice that $\sum_{r \in pos(\{r_1, r_2\}, y)} f_w(r, c_1) = f_w(r_1, c_1) = 5$, and that $\sum_{r \in neg(\{r_1, r_2\}, y)} f_w(r, c_1) = f_w(r_2, c_1) = 5$. From this and Definition 4.5, we get $\mathcal{P}^R(c_1, w_\mathcal{C}) = 0$. For Sensitivity to be satisfied, we would have to have $\mathcal{P}^R(c_1 \cup \{r\}, w_\mathcal{C}) = +$ for every context $c_1 \cup \{r\}$ where $r = (x, y, +)$. Notice that $\sum_{r \in pos(\{r_1, r_2, r_3\}, y)} f_w(r, c_2) = f_w(r_1, c_2) + f_2(r_3, c_2) = 5 + 0.5 = 5.5$, and

---

[10] The only exception is Absolute Weight Satisficing which (vacuously) satisfies Sensitivity.

	Add	Mul	Max	(Add)Thr	RelSat	AbsSat
3.5 Re	✓	✓	✓	✓	✓	✓
3.6 Ne	-	-	-	-	-	-
5.1 PoMn	✓	-	✓	✓	✓	✓
5.2 CoRe	✓	✓	-	✓	-	-
5.3 Se	✓	-	-	-	-	✓
5.4 UnMn	-	-	✓	-	-	-

Table 1
Summary of the principle-based analysis, assuming Fixed Weight

that $\sum_{r \in neg(\{r_1, r_2, r_3\}, y)} f_w(r, c_2) = f_w(r_2, c_2) = 5$. From this and Definition 4.5, we have $\mathcal{P}^R(c_2, w_\mathcal{C}) = 0$. $\qquad\square$

Proposition 5.6 *Maximizing balancing (Definition 4.3) with Fixed Weight (Principle 3.7) satisfies Polarity Monotony (Principle 5.1).*

Proof. Let $\mathcal{B}_o = (\mathcal{C}, w_\mathcal{C}, \mathcal{P}^m)$ be a maximizing balancing operator with a fixed weight system. Consider an arbitrary context $c = (R, y) \in \mathcal{C}$ such that $\mathcal{P}^m((R, y), w_\mathcal{C}) = v$ and $v \neq 0$. Assume that there is some reason $r' = (x, y, v)$ and a context $(R \cup \{r'\}, y) \in \mathcal{C}$. To establish that Polarity Monotony is satisfied, we need to show that $\mathcal{P}^m((R \cup \{r'\}, y), w_\mathcal{C}) = v$. Without loss of generality, we assume that $v = +$. From $\mathcal{P}^m((R, y), w_\mathcal{C}) = v$ and Definition 4.3, we know that $\max(\{f_w(r, (R, y)) \mid r \in pos(R, y)\}) = P > N = \max(\{f_w(r, (R, y)) \mid r \in neg(R, y)\})$. Now notice that $f_w(r', (R \cup \{r'\}, y)) > 0$, and that in $(R \cup \{r'\}, y)$ reasons have the same weights that they had in (R, y). From here, $\max(\{f_w(x, (R \cup \{r'\}, y)) \mid r \in pos(R, y) \cup \{r'\}\}) = \max(P, f_w(r', (R \cup \{r'\}, y))) \geq P > N = \max(\{f_w(r, (R, y)) \mid r \in neg(R, y)\})$. And this is enough to conclude that $\mathcal{P}^m((R \cup \{r\}, y), w_\mathcal{C}) = +$. $\qquad\square$

The proofs of other propositions—that is, the propositions that show which of the remaining operators do (or do not) satisfy which propositions—are equally straightforward. We leave them for a technical report and let Table 1 summarize the results that they establish: the topmost row lists the balancing operators; the leftmost column lists the principles; the remaining cells state whether the given operator does (✓) or doesn't (−) satisfy the given principle. For example, the third column makes it clear that the class of multiplicative operators (this is what Mul stand for) satisfy only two principles, namely, Relevance (Rel) and Commensurate Removal (CoRe).

It may be surprising to see that none of the operators satisfy Neutrality. Recall that Knoks and van der Torre [8] thought that both Relevance and Neutrality formalize properties that seem to be inherent in the metaphor of weighing reasons on scales. It turns out that the operators do not, in general, satisfy Neutrality with the assumption of Fixed Weight for the same reason that they do not, in general, satisfy all other principles without the assumption of Fixed Weight: nothing in the definition of fixed weight systems precludes them

43

from assigning $(x, y, +)$ and $(x, y, -)$ different weights in different contexts.

We can, of course, define a notion in the vicinity of fixed weight systems that makes this impossible.

Definition 5.7 [Symmetric weight systems] Let $\mathcal{C}$ be a set of contexts and $w_\mathcal{C} = (W, f_w)$ a weight system for $\mathcal{C}$. Then $w_\mathcal{C}$ is called a *symmetric weight system* just in case, for any pair of reasons $r = (x, y, +)$, $r' = (x, y, -)$ and any pair of contexts $c, c' \in \mathcal{C}$, we have $f_w(r, c) = f_w(r', c')$.

The counterpart of Fixed Weight then runs thus:

Principle 5.8 (Symmetry) *A balancing operator* $\mathcal{B}_o = (\mathcal{C}, w_\mathcal{C}, \mathcal{P})$ *satisfies* Symmetry, ***Sym***, *just in case (i)* $(R, y) \in \mathcal{C}$ *if and only if* $(\{(x, y, \overline{v}) : (x, y, v) \in R\}, y) \in \mathcal{C}$ *and (ii)* $w_\mathcal{C}$ *is a symmetric weight system.*

It is not difficult to verify that Symmetry entails Fixed Weight. What's more, it turns out that, with Symmetry in the background, every balancing operator from Section 4 satisfies Neutrality. Here is a sample proof.

Proposition 5.9 *Additive balancing (Definition 4.1) with Symmetry (Principle 5.8) satisfies Neutrality (Principle 3.6).*

Proof. Consider some additive operator $\mathcal{B}_o = (\mathcal{C}, w_\mathcal{C}, \mathcal{P}^+)$ that satisfies Symmetry. Consider an arbitrary context $c = (R, y)$ for which we have $\mathcal{P}^+((R, y), w_\mathcal{C}) = v$. Assume that $(R', y) \in \mathcal{C}$ where $R' = \{(x, y, \overline{v})\}$. Without loss of generality, suppose that $v = +$. Then we know that $\sum_{r \in pos(R,y)} f_w(r, (R, y)) = P > N = \sum_{r \in neg(R,y)} f_w(r, (R, y))$. Since $w_\mathcal{C}$ is symmetric, we know that $f_w((x, y, \overline{v}), (R', y)) = f_w((x, y, v), (R, y))$ for every $(x, y, \overline{v}) \in R'$. As a consequence, $\sum_{r \in pos(R', y)} f_w(r, (R', y)) = N < P = \sum_{r \in neg(R', y)} f_w(r, (R', y))$, and, thus, $\mathcal{P}^+((R, y), w_\mathcal{C}) = -$. $\square$

6 Related work

In this section, we relate our extension of Knoks and van der Torre's framework to the original. First of all, it's worth emphasizing that the original framework with its notion of a detachment systems is more general. For instance, it does not assume that the relation between contexts and values is functional, nor makes any restrictions on the shape of the set of values $\mathcal{V}$. Our notion of balancing operators is grounded in some conceptual choices and so it is more specific. Nevertheless, because of those conceptual choices it is also better suited to capture the informal model of weighing reasons on weight scales from the philosophical literature.

Knoks and van der Torre discuss two different classes of balancing *operations* (which qualify as specific types of detachment systems): what they call *anonymous* and *relational balancing operations*. It wouldn't be difficult to restate all the particular anonymous operations that they define as balancing operators. [11] For instance, Knoks and van der Torre's *Simple Counting* assigns a value to a context of the form (R, y) by comparing the number of positive and negative

---

[11] This, again, speaks to the flexibility of the formal notion of a balancing operator.

y-reasons in R. In the present framework, Simple Counting corresponds to a special case of Additive Balancing, namely, one the underlying weight system of which assigns the same weight to all reasons. (We leave the proof for the journal version of this paper.) Other anonymous balancing operations are straightforward to redefine as operators. Knoks and van der Torre formulated several principles that were satisfied by all of their anonymous balancing operations, but that do not hold for every operator we defined in Section 4. This includes Relevance and Polarity Monotony. As we saw, Relevance does not hold for all of these operators *unless* Fixed Weight is also assumed, and Polarity Monotony does not hold for Multiplicative Balancing *even* if Fixed Weight is assumed. These observations illustrate that the notion of a balancing operator gives us a grip on richer structures.

Turning to relational balancing operations, these are more difficult to relate to balancing operators, since relational operations come equipped with a relation over reasons. It turns out to be possible to establish a connection between Maximizing Balancing Operators and one particular relational operation: what Knoks and van der Torre call *Decisive Reason*. This operation assigns a value to a context by checking the polarity of the "strongest" reason in it. It shouldn't be difficult to see that the "stronger than" relation can be mapped to the greater than relation of numerical weights, and that, with this mapping, Decisive Reasons has the form of Maximizing Balancing operators. (Again, we leave the proof of this for the journal version.) To be in a position to explore the connections between relational and numerical balancing, it would pay to extend the notion of a balancing operator with a further component, namely, a binary anti-symmetric relation over reasons (in contexts). With this, we would be in a position to formulate principles that have to do with the relation—in addition to principles that have to do with weight systems and procedures—and have a general framework for analyzing anonymous, numerical, and rational balancing, as well as the connections between them.

7 Conclusion and future work

In this paper, we extended Knoks and van der Torre's framework [8] to richer structures in which reasons are associated with numerical weights. We started by introducing the formal notions of weight systems, procedures, and balancing operators. Then we introduce six concrete classes of balancing operators, presented a principle-based analysis of them, and explained how the results presented here go beyond those of [8].

For future work, we plan to set up and explore the more general framework we mentioned at the end of previous section, that is, a framework that would unify numerical and relational balancing. It seems to be clear that what we have done above shows that there is a rich variety of balancing operators available for exploration and formal analysis. We also plan to explore how our numerical balancing framework relates to multi-criteria decision-making [7] and qualitative bipolar decision-making [4], as well as how it might be used to model case-based reasoning.

References

[1] Berker, S., *Particular reasons*, Ethics **118** (2007), pp. 109–139.

[2] Broome, J., *Reasons*, in: R. J. Wallace, P. Petti, S. Scheffler and M. Smith, editors, *Reasons and Value: Themes from the Moral Philosophy of Joseph Raz*, Oxford: Clarendon Press, 2004 pp. 28–55.

[3] Dietrich, F. and C. List, *A reason-based theory of rational choice*, Noûs **47** (2013), pp. 104–34.

[4] Dubois, D., G. Fargier and J.-L. Bonnefon, *On the qualitative comparison of decisions having positive and negative features*, Journal of Artificial Intelligence Research **32** (2008), pp. 385–417.

[5] Horty, J., *Reasons as defaults*, Philosophers' Imprint **7** (2007), pp. 1–28.

[6] Horty, J., "Reasons as Defaults," Oxford University Press, 2012.

[7] Keeney, R. and H. Raiffa, "Decisions with Multiple Objectives: Preferences and Value," Cambridge University Press, 1993.

[8] Knoks, A. and L. van der Torre, *Reason-based detachment*, in: B. Bentzen, B. Liao, D. Liga, R. Markovich, B. Wei, M. Xiong and T. Xu, editors, *Joint Proceedings of the 3rd International Workshop on Logics for New-Generation Artificial Intelligence and the International Workshop on Logic, AI, and Law (LNGAI/LAIL2023, Hangzhou)* (2023), pp. 49–65.

[9] Lord, E. and B. Maguire, *An opinionated guide to the weight of reasons*, in: E. Lord and B. Maguire, editors, *Weighing Reasons*, Oxford University Press, 2016 pp. 3–24.

[10] Nair, S., *"Adding up" reasons: Lessons for reductive and non-reductive approaches*, Ethics **132** (2021), pp. 38–88.

[11] Parfit, D., "On What Matters (Volume I)," Oxford University Press, 2011.

[12] Raz, J., "Practical reason and norms," Oxford University Press, 1990.

[13] Scanlon, T. M., "What We Owe to Each Other," Cambridge, MA: Harvard University Press, 1998.

[14] Sher, I., *Comparative value and the weight of reasons*, Economics and philosophy **35** (2019), pp. 103–158.

[15] Snedegar, J., *Reasons for and reasons against*, Philosophical Studies **175** (2018), pp. 725–43.

[16] Tucker, C., *The dual scale model of weighing reasons*, Noûs **56** (2022), pp. 366–92.

[17] Tucker, C., *A holist balance scale*, Journal of the American Philosophical Association **9** (2023), pp. 533–53.

[18] Tucker, C., "The Weight of Reasons: A Framework for Ethics," Oxford University Press, forthcoming.

[19] Whiting, D., "The Range of Reasons: in Ethics and Epistemology," Oxford University Press, 2021.

Towards a Logical Approach to Recommendations

Fenrong Liu, Wei Wang, and Sisi Yang

The Tsinghua-UvA JRC for Logic, Department of Philosophy, Tsinghua University

Abstract

In the digital era, users encounter an endless stream of recommendations. The development of recommendation systems in AI has attracted extensive attention, yielding a substantial body of literature. However, we have not encountered any logical systems for reasoning about recommendation systems, despite the immense amount of reasoning involved in making recommendations. In this paper, we propose a new recommendation logic (**RL**) to study the reasoning behind recommendations, emphasizing their basis in users' revealed preferences. We explore the expressivity of **RL** by introducing a new notion of bisimulation and translating **RL** into a 3-variable fragment of a two-sorted first-order logic. We show that **RL** has the tree model property and that its model-checking problem can be solved in polynomial time, for which we propose an algorithm and prove its correctness. We believe that our work has the potential to advance personalized recommendations.

Keywords: recommendation, preference, modal logic, model checking.

1 Introduction: logic for data-based recommendations

In our increasingly digital world, we are constantly bombarded with recommendations generated by recommendation systems, from movie suggestions on streaming services to product endorsements online. To date, significant attention has been focused on the *algorithms* that play a crucial role in filtering and ranking these recommendations, resulting in an ever-growing body of literature. [15] is a widely cited reference that includes various major algorithms for recommendations. More recent surveys can be found in [9] and [16]. This field also intersects with data mining (see, for example, [6], [11]), aiming to discern customers' preferences.

Logical methods (e.g., fuzzy logic [8], [14], [16]) have been applied in the field of recommendation systems. Some logical studies on the foundations of recommendation systems have also been conducted, including logical formalization of filtering conditions ([12], [2]) and knowledge-based recommendation systems ([4]). Nevertheless, we have not seen any logical system for reasoning about recommendations made by recommendation systems, despite the fact that making recommendations based on data inherently involves substantial reasoning and

47

decision-making. This paper aims to fill the gap. Our approach is founded on the explicit assumption that recommendations ought to be grounded in data reflecting users' choices. This leads us to embrace the theory of *revealed preference* from economics (e.g., [17], [19]), which suggests that by analyzing users' actions—such as interactions, selections, or purchases—we can infer their preferences, even in the absence of explicit ratings. Our *aim* is to uncover the underlying logic that connects the vast amount of data with the personalized recommendations users receive, enabling us to reason about recommendation algorithms. Such a logical characterization would equip us with an abstract view of algorithms, and thus has the potential to provide insights on how to enhance existing algorithms and inspire new ones.

Let us consider a realistic example:

Example 1.1 [Shopping on three platforms] Alice has been shopping on three e-commerce platforms: Taobao (S_1), Pinduoduo (S_2), and JD.com (S_3), as shown in Figure 1. We denote statements such as "Alice bought a box of copy paper," "... a toner cartridge," "... a printer," and "... a stapler" by propositional letters p_1, p_2, p_3, and p_4, respectively. Data point d_1 shows that in Jan. 2023, Alice bought a box of copy paper and a toner cartridge on S_1. In May 2023, she purchased boxes of copy paper on both S_2 and S_3; in addition, she bought a printer along with a toner cartridge on S_3 (possibly because her old printer was broken), as shown by data points d_2 and d_3. Later in May, d_4 recorded purchases of a box of copy paper and a stapler, not tracked by any platforms. Finally, d_5 indicates that in Sep. 2023, Alice bought boxes of copy paper and toner cartridges on both S_1 and S_2.[1]

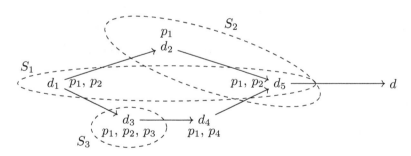

Fig. 1. Purchase on three platforms

Given Alice's purchase data, what can a recommendation system running at d learn about Alice's preferences, and based on that, how can it make personalized recommendations?

Let us first see what we can observe or say at d:

---

[1] Appendix A includes our remarks on the branching structure in Figure 1.

(i) What Alice **kept doing** on **every platform**: p_1 occurs in all data points before d that are collected by some platform. Naturally, a recommendation system should *strongly recommend* copy paper to her.

(ii) What Alice **has done** on **every platform**: On all platforms there is a data point before d where p_2 is true, indicating Alice bought toner cartridges on each platform. This consistency constitutes a *good reason to recommend* toner cartridges, though the basis for recommendation is not as strong as that in case (i).

(iii) What Alice **kept doing** on **some platform**: p_3 occurs in all data points of platform S_3, whereas $\neg p_3$ holds at all data points of S_1 and S_2, indicating potential *platform preferences* of Alice for printer purchases. Thus, recommendations can be tailored based on data from specific platforms. [2]

(iv) What Alice **has done** on **some platform**: This constitutes the minimal requirement for recommendations based on direct evidence.

Note the combination of quantifiers on platforms and data points in the above four notions. The implications of this combination will be studied in this paper, where we will formalize those notions of recommendations.

It is noteworthy that the above example can be easily adapted into a typical scenario of Content-Based Filtering ([15], [9], [16]), a major type of recommendation models that recommends items similar to those selected by the user in the past. The point is that propositional letters can also represent attributes of items, so that we can reason about whether items with certain attributes should be recommended to the user given the attributes of the items she purchased. For instance, in the above example, we can use propositional letters to denote attributes such as "Alice bought a printer-related item," "... office supplies," "... an item in the price range of 200–500 yuan," etc.

The paper is structured as follows: A new formal language and semantics for recommendations are introduced in Sec. 2, which can be used to formalize Example 1.1 and hence the typical reasoning behind Content-Based Filtering. A novel bisimulation concept is discussed in Sec. 3.1. The standard translation into two-sorted first-order logic is given in Sec. 3.2. A series of computational results are shown in Sec. 4. Translations between our models and purely Kripke or neighborhood models are studied in Sec. 5. The paper concludes in Sec. 6.

2 RL: language and semantics

In what follows, we will introduce a new recommendation logic (**RL**), and present its language and semantics.

Definition 2.1 [Language] Let **Prop** be a countably infinite set of propositional letters. The **RL**-*language* $\mathcal{L}$ is defined as follows:

$$\varphi ::= p \in \textbf{Prop} \mid \bot \mid \neg\varphi \mid \varphi \wedge \varphi \mid [R]\varphi \mid \langle R \rangle\varphi$$

---

[2] We are aware of the existing literature focusing on preferences or selection between different social media platforms ([5], [7], [10], [20]), primarily from the perspective of algorithms.

This language extends propositional logic with two new modalities, $[R]$ and $\langle R]$. Their duals will be written as $\langle R \rangle$ and $[R\rangle$, respectively.[3]

To give a precise interpretation of these operators, we define **RL**-models.

Definition 2.2 [Frames and models] An **RL**-*frame* is defined as a triple $(D, \{S_i\}_{i \in I}, \prec)$, where

(i) D is a set of *data points*.

(ii) $\{S_i\}_{i \in I}$ is a non-empty set of *data sequences*[4] with each $S_i \subseteq D$.

(iii) $\prec$ is a *precedence relation* between data points which is irreflexive and transitive (i.e., is a strict partial order).

An **RL**-*model* is a tuple $(D, \{S_i\}_{i \in I}, \prec, V)$, where $(D, \{S_i\}_{i \in I}, \prec)$ is an **RL**-frame and V is a *valuation* that assigns to every propositional letter p the set of data points $V(p) \subseteq D$ at which p is true.

Fig. 1 in Example 1.1 illustrates an **RL**-model, which is built on data points, each detailing the behavior of the user (e.g., clicked at what, bought what, etc.). Some of these data points are collected by specific platforms, whereas others might not be collected by any platform.

Definition 2.3 [Truth] The *truth* of an arbitrary formula $\varphi \in \mathcal{L}$ in an **RL**-model $\mathfrak{M} = (D, \{S_i\}_{i \in I}, \prec, V)$ at $d \in D$ is defined inductively as follows:

$$
\begin{array}{lll}
\mathfrak{M}, d \models p & \text{iff} & d \in V(p) \\
\mathfrak{M}, d \models \bot & \text{iff} & \text{never} \\
\mathfrak{M}, d \models \neg\varphi & \text{iff} & \mathfrak{M}, d \not\models \varphi \\
\mathfrak{M}, d \models \varphi \wedge \psi & \text{iff} & \mathfrak{M}, d \models \varphi \text{ and } \mathfrak{M}, d \models \psi \\
\mathfrak{M}, d \models [R]\varphi & \text{iff} & \text{for all } i \in I \text{ and all } d' \in S_i \text{ with } d' \prec d, \text{ we have} \\
& & \mathfrak{M}, d' \models \varphi \\
\mathfrak{M}, d \models \langle R]\varphi & \text{iff} & \text{there exists } i \in I \text{ s.t. for all } d' \in S_i \text{ with } d' \prec d, \\
& & \text{we have } \mathfrak{M}, d' \models \varphi
\end{array}
$$

The operators $[R]$, $[R\rangle$, $\langle R]$, and $\langle R \rangle$ formalize the four notions of recommendations (i)–(iv) discussed in Sec. 1, respectively, representing varying recommendation strength or ranks. Specifically, $[R]$ and $\langle R \rangle$ indicate strong and weak recommendations, respectively. $[R\rangle$ and $\langle R]$ denote intermediate strength, focusing on cross-platform and platform-specific preferences, respectively.

Readers may find that our semantics is a mixture of Kripke semantics and neighborhood semantics. A close comparison with these semantics will be addressed in Section 5, where one can see that **RL**-models are more intuitive and compact than their purely Kripke or neighborhood counterparts.

---

[3] The brackets in those operators have very intuitive meanings: left (right) brackets represent quantifiers on platforms (data points), and square (angle) brackets represent universal (existential) quantifiers. E.g., [denotes "on every platform."

[4] Or informally *platforms*, identifying data sequences with platforms that provide them.

Finally, validity of an $\mathcal{L}$-formula φ (notation $\models \varphi$) is defined as usual. To help readers gain a better understanding of our semantics, we present some interesting (in)validities.

Fact 2.4 *The following formulas are valid and demonstrate the relative strength of the recommendation operators.*

$$\models [R]\varphi \to \langle R\rangle\varphi \qquad \models [R\rangle\varphi \to \langle R\rangle\varphi$$

Fact 2.5 *The following formulas are valid (invalid).* [5]

(K) $\models [R](\varphi \to \psi) \to ([R]\varphi \to [R]\psi)$
           $\not\models \langle R\rangle(\varphi \to \psi) \to (\langle R\rangle\varphi \to \langle R\rangle\psi)$

(M) $\models \langle R\rangle(\varphi \wedge \psi) \to \langle R\rangle\varphi \wedge \langle R\rangle\psi$

(C) $\not\models \langle R\rangle\varphi \wedge \langle R\rangle\psi \to \langle R\rangle(\varphi \wedge \psi)$

(N) $\models \langle R\rangle\top$

(4) $\models [R]\varphi \to [R][R]\varphi$
        $\models \langle R\rangle\varphi \to \langle R\rangle\langle R\rangle\varphi$

(Int) $\models [R](\varphi \to \psi) \to (\langle R\rangle\varphi \to \langle R\rangle\psi)$
        $\models \langle R\rangle\varphi \to [R]\langle R\rangle\varphi$

3   Expressive power

The expressivity of $\mathcal{L}$ is studied in two respects: Sec. 3.1 explores it structurally via bisimulation, while Sec. 3.2 compares $\mathcal{L}$ with first-order language, showing that **RL** is equivalent to a 3-variable fragment of first-order logic through the standard translation.

3.1   Bisimulation

In what follows, we propose a new notion of bisimulation for **RL**-models, which captures the adequate notion of modal equivalence.

Definition 3.1 [Bisimulation] A *bisimulation* between two **RL**-models $\mathfrak{M} = (D, \{S_i\}_{i\in I}, \prec, V)$ and $\mathfrak{M}' = (D', \{S_i'\}_{i\in I'}, \prec', V')$ is a non-empty relation $Z \subseteq D \times D'$ such that dZd' iff

(i) (*atomic* condition): For each $p \in \mathbf{Prop}$, $d \in V(p)$ iff $d' \in V'(p)$.

(ii) (*forth* condition):
     (a) For each $e \in \bigcup_{i\in I} S_i$ such that $e \prec d$, there exists $e' \in \bigcup_{i\in I'} S_i'$ such that $e' \prec' d'$ and eZe'.
     (b) For each S_i, there exists S_j' such that for each $e' \in S_j'$ with $e' \prec' d'$, there exists $e \in S_i$ such that $e \prec d$ and eZe'.

(iii) (*back* condition): Similar to the forth condition.

---

[5] [13] shows that the logic **K** is equivalent to the smallest minimal modal logic $\mathbf{E} + (M) + (C) + (N)$. In light of this and the following results, the invalidity of (C) for $\langle R\rangle$ explains the invalidity of (K) for $\langle R\rangle$, and thus why $\langle R\rangle$ is not a normal modal operator.

When there is a bisimulation Z linking $d \in \mathfrak{M}$ and $d' \in \mathfrak{M}'$, we say that d and d' are *bisimilar*, notation $Z : \mathfrak{M}, d \leftrightarrow \mathfrak{M}', d'$ or simply $\mathfrak{M}, d \leftrightarrow \mathfrak{M}', d'$.

Basically, two points linked by a bisimulation share the same atomic information (the atomic condition) and have corresponding transition possibilities (the forth and back conditions). The forth and back conditions reflect the nature of **RL**-models as a mixture of relational models and neighborhood models: (a) comes from the standard bisimulation for relational semantics ([3], [18]), whereas (b) is inspired by bisimulations for neighborhood semantics ([13], [18]).

We then show that bisimulations thus defined characterize modal equivalence on **RL**-models in the following sense: bisimilarity implies modal equivalence (Theorem 3.3), while the converse holds on image-finite **RL**-models (Theorem 3.5).

Definition 3.2 [Modal equivalence] Given two **RL**-models $\mathfrak{M} = (D, \{S_i\}_{i \in I}, \prec, V)$ and $\mathfrak{M}' = (D', \{S'_i\}_{i \in I'}, \prec', V')$, $d \in D$, and $d' \in D'$, we say $\mathfrak{M}, d$ and $\mathfrak{M}', d'$ are *modally equivalent* (notation $\mathfrak{M}, d \equiv \mathfrak{M}', d'$), iff

$$\text{for each } \varphi \in \mathcal{L},\ \mathfrak{M}, d \models \varphi \text{ iff } \mathfrak{M}', d' \models \varphi$$

Theorem 3.3 *Let* $\mathfrak{M} = (D, \{S_i\}_{i \in I}, \prec, V)$ *and* $\mathfrak{M}' = (D', \{S'_i\}_{i \in I'}, \prec', V')$ *be* **RL***-models. For each* $d \in D$ *and* $d' \in D'$, $Z : \mathfrak{M}, d \leftrightarrow \mathfrak{M}', d'$ *implies* $\mathfrak{M}, d \equiv \mathfrak{M}', d'$.

Proof. We refer to the proof in the Appendix. $\square$

As in relational semantics, the other direction does not hold generally, but for *image-finite* **RL**-models, bisimilarity does imply modal equivalence.

Definition 3.4 [Image-finiteness] Given a relation $R \subseteq X \times Y$ and $x \in X$, the *image* $R[x]$ of x under R is the set $\{y \in Y : xRy\}$.

An **RL**-model $\mathfrak{M} = (D, \{S_i\}_{i \in I}, \prec, V)$ is *image-finite*, iff for each $d \in D$, $\prec^{-1}[d] \cap \bigcup_{i \in I} S_i$ and I are finite.

Theorem 3.5 *Let* $\mathfrak{M} = (D, \{S_i\}_{i \in I}, \prec, V)$ *and* $\mathfrak{M}' = (D', \{S'_i\}_{i \in I'}, \prec', V')$ *be image-finite* **RL***-models. For each* $d \in D$ *and* $d' \in D'$, $\mathfrak{M}, d \equiv \mathfrak{M}', d'$ *implies* $\mathfrak{M}, d \leftrightarrow \mathfrak{M}', d'$.

Proof. See the proof in the Appendix. $\square$

3.2 The standard translation

We are about to show that every $\mathcal{L}$-formula is equivalent to a two-sorted first-order formula with at most three variables. We first define the correspondence language into which we will translate $\mathcal{L}$-formulas.

Definition 3.6 [Correspondence language] The *correspondence language* $\mathcal{L}_{\text{fo}}$ is a two-sorted first-order language whose two sorts are d and s, intended to represent data points and data sequences, respectively. d consists of two variables x and y, while s consists of only one variable s.

The signature of $\mathcal{L}_{\text{fo}}$ contains unary predicates $P_0, P_1, \ldots$ of sort d for each propositional letter $p_0, p_1, \cdots \in \mathbf{Prop}$, a binary relation symbol R relating two elements of sort d, and a binary relation symbol E relating elements of sort d to elements of sort s. The intended interpretation of xRy is "x precedes y," and the intended interpretation of xEs is "x is an element of s."

$\mathcal{L}_{\mathrm{fo}}$ is generated by the following grammar:

$$x = y \mid P_i x \mid xRx \mid xEs \mid \neg\varphi \mid \varphi \wedge \varphi \mid \exists x\varphi \mid \exists s\varphi$$

We then define the first-order translation of **RL**-models on which we interpret $\mathcal{L}_{\mathrm{fo}}$-formulas.

Definition 3.7 [First-order translation of **RL**-models] Suppose that $\mathfrak{M} = (D, \{S_i\}_{i \in I}, \prec, V)$ is an **RL**-model. The *first-order translation of* $\mathfrak{M}$ is the first-order structure $\mathfrak{M}^* = (W, \{P_i\}_{i \in \mathbb{N}}, R, E)$ where:

(i) $W = W^d \cup W^s$ with $W^d = D$, $W^s = \{S_i\}_{i \in I}$.

(ii) $P_i = V(p_i)$ for each $p_i \in \mathbf{Prop}$.

(iii) $R = \{(d', d) : d', d \in W^d \text{ and } d' \prec d\}$.

(iv) $E = \{(d, S_i) : d \in W^d, \ S_i \in W^s, \text{ and } d \in S_i\}$.

Finally, we define the standard translation of $\mathcal{L}$ into $\mathcal{L}_{\mathrm{fo}}$.

Definition 3.8 (Standard translation) The *standard translation* of $\mathcal{L}$ consists of $ST_x : \mathcal{L} \to \mathcal{L}_{\mathrm{fo}}$ and $ST_y : \mathcal{L} \to \mathcal{L}_{\mathrm{fo}}$ defined by mutual induction:

$$ST_x(p) = Px$$
$$ST_x(\bot) = x \neq x$$
$$ST_x(\neg\varphi) = \neg ST_x(\varphi)$$
$$ST_x(\varphi \wedge \psi) = ST_x(\varphi) \wedge ST_x(\psi)$$
$$ST_x([R]\varphi) = \forall s \forall y (yEs \wedge yRx \to ST_y(\varphi))$$
$$ST_x(\langle R\rangle\varphi) = \exists s \forall y (yEs \wedge yRx \to ST_y(\varphi))$$

$$ST_y(p) = Py$$
$$ST_y(\bot) = y \neq y$$
$$ST_y(\neg\varphi) = \neg ST_y(\varphi)$$
$$ST_y(\varphi \wedge \psi) = ST_y(\varphi) \wedge ST_y(\psi)$$
$$ST_y([R]\varphi) = \forall s \forall x (xEs \wedge xRy \to ST_x(\varphi))$$
$$ST_y(\langle R\rangle\varphi) = \exists s \forall x (xEs \wedge xRy \to ST_x(\varphi))$$

We now prove the following theorem, which says that every $\mathcal{L}$-formula is equivalent to an $\mathcal{L}_{\mathrm{fo}}$-formula containing at most three variables.

Theorem 3.9 *Let* $\mathfrak{M} = (D, \{S_i\}_{i \in I}, \prec, V)$ *be an* **RL**-*model and* $\varphi \in \mathcal{L}$. *Then*

(i) *For each* $d \in D$, $\mathfrak{M}, d \models \varphi$ *iff* $\mathfrak{M}^* \models ST_x(\varphi)[d]$ *iff* $\mathfrak{M}^* \models ST_y(\varphi)[d]$.

(ii) $\mathfrak{M} \models \varphi$ *iff* $\mathfrak{M}^* \models \forall x ST_x(\varphi)$.

Proof. Again, we refer to the proof in the Appendix. □

4 Computational properties

In this section, we show that **RL** has two nice computational properties: it enjoys the tree model property (Sec. 4.1), which is usually a positive indicator for the computational behavior of a logic; its model-checking problem can be solved in polynomial time (Sec. 4.2), paving the way for its applications in recommendation systems.

4.1 Tree model property

We first explain what the tree model property is.

Definition 4.1 [Tree-like model] (T, R) is a (transitive) *tree*, iff

(i) $R \subseteq T \times T$.

(ii) There exists a *root* $r \in T$ such that Rrt for each $t \in T$ with $t \neq r$.

(iii) For each $t \in T$, $\{s \in T : Rst\}$ is finite and linearly ordered by R.

An **RL**-model $\mathfrak{M} = (D, \{S_i\}_{i \in I}, \prec, V)$ is *tree-like* iff $(D, \prec^{-1})$ is a tree.

Theorem 4.2 (Tree model property) **RL** *has the tree model property: any* **RL***-satisfiable* $\varphi \in \mathcal{L}$ *is satisfiable in a tree-like* **RL***-model.*

We prove the above theorem by a classical method called *unraveling*.

Definition 4.3 [Unraveling] Let $\sigma_1 + \sigma_2$ denote the concatenation of two sequences σ_1 and σ_2, and let $(\sigma)_0$ denote the first term of a sequence σ.

Given an **RL**-model $\mathfrak{M} = (D, \{S_i\}_{i \in I}, \prec, V)$ and $d \in D$, the *unraveling* of $\mathfrak{M}$ around d is the **RL**-model $\mathfrak{M}^u = (D^u, \{S_i^u\}_{i \in I}, \prec^u, V^u)$ where:

(i) D^u is the set of all finite sequences $(e_n, ..., e_1, d)$ such that $e_n \prec e_{n-1} \prec \cdots \prec e_1 \prec d$.

(ii) For each $i \in I$, $S_i^u = \{\sigma \in D^u : (\sigma)_0 \in S_i\}$.

(iii) $\prec^u = \{(\sigma', \sigma) \in D^u \times D^u : \text{there exists } d \in D \text{ such that } \sigma' = (d) + \sigma\}$.

(iv) For each $p \in \mathbf{Prop}$, $\sigma \in V^u(p)$ iff $(\sigma)_0 \in V(p)$.

Given a binary relation $R \subseteq S \times S$, the *transitive closure* of R is the smallest transitive relation on S that contains R.

It is easy to see that the following fact holds:

Fact 4.4 *Given an* **RL***-model* $\mathfrak{M} = (D, \{S_i\}_{i \in I}, \prec, V)$ *and* $d \in D$, *the transitive closure of the unraveling of* $\mathfrak{M}$ *around* d *is a tree-like* **RL***-model.*

Finally, a proof of the tree model property of **RL** (Theorem 4.2) in virtue of unraveling can be found in the Appendix.

4.2 Time complexity of model checking

Definition 4.5 [Model-checking problem] The **RL** *model-checking problem* is as follows: given a finite **RL**-model $\mathfrak{M} = (D, \{S_i\}_{i \in I}, \prec, V)$, $d \in D$, and $\varphi \in \mathcal{L}$, determine whether $\mathfrak{M}, d \models \varphi$.

Application: It is not difficult to see how **RL** model checking can be useful. For instance, consider again Example 1.1. Let $\mathfrak{M}$ denote the model illustrated by Figure 1. A recommendation system can determine by **RL** model checking which of $[R]p$, $\langle R \rangle p$, $[R\rangle p$, and $\langle R]p$ (for $p \in \{p_1, p_2, p_3, p_4\}$) hold at $\mathfrak{M}, d$, and determine its recommendation behavior (e.g., whether to recommend, the priority or rank of recommendation) accordingly.

As usual, we show that **RL** model checking can be done by first recursively calculating the satisfaction set for each subformula of φ by a bottom-up traversal over the parse tree of φ, then checking whether d belongs to the satisfaction set of φ (see [1]). Furthermore, the entire procedure can be performed in *polynomial time*, as indicated by the following theorem.

Theorem 4.6 *The* **RL** *model-checking problem can be solved in*

$$O(|I| \cdot (|D| + | \prec |) \cdot |\varphi|)$$

where $|\varphi|$ is defined inductively as follows:

(i) $|p| = 1$ *for each* $p \in$ **Prop**.

(ii) $|\bot| = 1$.

(iii) $| \star \psi | = |\psi| + 1$ *for each unary operator* $\star$.

(iv) $|\psi_1 \circ \psi_2| = |\psi_1| + |\psi_2| + 1$ *for each binary operator* $\circ$.

In Sec. 4.2.1, we give a model-checking algorithm for **RL** and prove its correctness; in Sec. 4.2.2, we calculate the time complexity of this algorithm, which completes the proof of the above theorem.

4.2.1 A model-checking algorithm

The following concept is useful in developing an efficient model-checking algorithm.

Definition 4.7 [Satisfaction set] Given an **RL**-model $\mathfrak{M} = (D, \{S_i\}_{i \in I}, \prec, V)$ and $\varphi \in \mathcal{L}$, the *satisfaction set* of φ in $\mathfrak{M}$ is defined as

$$Sat^{\mathfrak{M}}(\varphi) = \{d \in D : \mathfrak{M}, d \models \varphi\}$$

A model-checking algorithm is given by the function CHECK in Algorithm 1, which calls the function SAT to compute satisfaction sets. The correctness of those two functions is guaranteed by the following lemma.

Lemma 4.8 *For each finite* **RL**-*model* $\mathfrak{M} = (W, \{S_i\}_{i \in D}, \prec, V)$, $d \in D$ *and* $\varphi \in \mathcal{L}$,

(i) $Sat^{\mathfrak{M}}(\varphi) = $ SAT$(\mathfrak{M}, \varphi)$.

(ii) $\mathfrak{M}, d \models \varphi$ *iff* CHECK$(\mathfrak{M}, d, \varphi)$.

Proof. We refer to the proof in the Appendix. □

4.2.2 Time complexity of Algorithm 1

In the Appendix, we prove that the time complexity of Algorithm 1 is in $O(|I| \cdot (|D| + | \prec |) \cdot |\varphi|)$ (Theorem 4.6).

Algorithm 1. An **RL** model-checking algorithm
Require: Finite **RL**-model $\mathfrak{M} = (D, \{S_i\}_{i \in I}, \prec, V)$, $d \in D$, and $\varphi \in \mathcal{L}$
1: **function** SAT($\mathfrak{M}, \varphi$)   ▷ Computes $Sat^{\mathfrak{M}}(\varphi)$
2:     **if** $\varphi \in \mathbf{Prop}$ **then**
3:        **return** $V(\varphi)$
4:     **else if** $\varphi = \bot$ **then**
5:        **return** $\emptyset$
6:     **else if** $\varphi = \neg\psi$ **then**
7:        **return** $D\backslash$SAT($\mathfrak{M}, \psi$)
8:     **else if** $\varphi = \psi_1 \wedge \psi_2$ **then**
9:        **return** SAT($\mathfrak{M}, \psi_1$)$\cap$SAT($\mathfrak{M}, \psi_2$)
10:     **else if** $\varphi = \langle R\rangle\psi$ **then**
11:        $S \leftarrow \emptyset$
12:        **for all** $i \in I$ **do**
13:           $S \leftarrow S \cup S_i$
14:        **end for**
15:        $T \leftarrow \emptyset$
16:        **for all** $d \in$ SAT($\mathfrak{M}, \psi$)$\cap S$ **do**
17:           $T \leftarrow T\cup \prec[d]$
18:        **end for**
19:        **return** T
20:     **else if** $\varphi = [R\rangle\psi$ **then**
21:        $G \leftarrow$ SAT($\mathfrak{M}, \psi$)
22:        $T \leftarrow D$
23:        **for all** $i \in I$ **do**
24:           $E \leftarrow \emptyset$
25:           **for all** $d \in G \cap S_i$ **do**
26:              $E \leftarrow E\cup \prec[d]$
27:           **end for**
28:           $T \leftarrow T \cap E$
29:        **end for**
30:        **return** T
31:     **end if**
32: **end function**
33: **function** CHECK($\mathfrak{M}, d, \varphi$)   ▷ Checks if $\mathfrak{M}, d \models \varphi$
34:     **return** $d \in$SAT($\mathfrak{M}, \varphi$)
35: **end function**

5   Comparison

As we commented several times, **RL**-models are a mixture of Kripke models and neighborhood models. We will elaborate on the connections between **RL**-models and purely Kripke or neighborhood models. Readers will find **RL**-models more intuitive and compact than their purely Kripke or neighborhood counterparts.

5.1 Equivalence between RL-models and Kripke RL-models

In this part, we show the equivalence between **RL**-models and their purely Kripke counterparts – Kripke **RL**-models, in the sense that each **RL**-model can be translated into a modally equivalent Kripke **RL**-model, and vice versa.

Definition 5.1 [Kripke **RL**-model] A *Kripke **RL**-model* is defined as a triple $(D, \{\prec_i\}_{i \in I \cup \{I\}}, V)$, where

(i) $I \neq \emptyset$.

(ii) $\prec_i \subseteq D \times D$ for each $i \in I \cup \{I\}$.

(iii) $\bigcup_{i \in I \cup \{I\}} \prec_i$ is irreflexive and transitive.

(iv) *Forward consistency*: for each $d \in D$ and $i, j \in I \cup \{I\}$, if $\prec_i [d] \neq \emptyset$ and $\prec_j [d] \neq \emptyset$, then $\prec_i [d] = \prec_j [d]$.

(v) $V : \textbf{Prop} \to \mathcal{P}(D)$ is a *valuation*.

Definition 5.2 [Truth] The *truth* of an arbitrary formula $\varphi \in \mathcal{L}$ in a Kripke **RL**-model $\mathfrak{K} = (D, \{\prec_i\}_{i \in I \cup \{I\}}, V)$ at $d \in D$ is defined inductively as follows:

$$
\begin{array}{lll}
\mathfrak{K}, d \models p & \text{iff} & d \in V(p) \\
\mathfrak{K}, d \models \bot & \text{iff} & \text{never} \\
\mathfrak{K}, d \models \neg\varphi & \text{iff} & \mathfrak{M}, d \nvDash \varphi \\
\mathfrak{K}, d \models \varphi \wedge \psi & \text{iff} & \mathfrak{K}, d \models \varphi \text{ and } \mathfrak{K}, d \models \psi \\
\mathfrak{K}, d \models [R]\varphi & \text{iff} & \text{for all } i \in I \text{ and all } d' \in D \text{ with } d' \prec_i d, \text{ we have} \\
& & \mathfrak{K}, d' \models \varphi \\
\mathfrak{K}, d \models \langle R \rangle \varphi & \text{iff} & \text{there exists } i \in I \text{ such that for all } d' \in D \text{ with} \\
& & d' \prec_i d, \text{ we have } \mathfrak{K}, d' \models \varphi
\end{array}
$$

Fact 5.3 *Let* $\mathfrak{M} = (D, \{S_i\}_{i \in I}, \prec, V)$ *be an* **RL**-*model, and* $\mathfrak{K}^M = (D, \{\prec_i\}_{i \in I \cup \{I\}}, V)$, *where*

(i) *For each* $i \in I$,
$$\prec_i = \{(d', d) \in \prec : d' \in S_i\}$$

(ii) $\prec_I = \{(d', d) \in \prec : d' \notin \bigcup_{i \in I} S_i\}$.

$\mathfrak{K}^M$ *is a Kripke* **RL**-*model, called the* Kripke translation *of* $\mathfrak{M}$.

Proof. A proof is given in the Appendix. □

Theorem 5.4 *Let* $\mathfrak{M} = (D, \{S_i\}_{i \in I}, \prec, V)$ *be an* **RL**-*model, and* $\mathfrak{K}^M$ *be the Kripke translation of* $\mathfrak{M}$. *For each* $d \in D$, $\mathfrak{M}, d \equiv \mathfrak{K}^M, d$.

Proof. By induction on φ. □

We then introduce the translation in the other direction.

Definition 5.5 [**RL**-translation of Kripke **RL**-models] Given a Kripke **RL**-model $\mathfrak{K} = (D, \{\prec_i\}_{i \in I \cup \{I\}}, V)$, the **RL**-*translation of* $\mathfrak{K}$ is the **RL**-model $\mathfrak{M}^K = (D, \{S_i\}_{i \in I}, \prec, V)$, where:

(i) $S_i = \{d : \prec_i [d] \neq \emptyset\}$ for each $i \in I$.

(ii) $\prec = \bigcup_{i \in I \cup \{I\}} \prec_i$.

Theorem 5.6 *Let* $\mathfrak{K} = (D, \{\prec_i\}_{i \in I \cup \{I\}}, V)$ *be a Kripke* **RL**-*model, and* $\mathfrak{M}^K = (D, \{S_i\}_{i \in I}, \prec, V)$ *be the* **RL**-*translation of* $\mathfrak{K}$. *For each* $d \in D$, $\mathfrak{K}, d \models \varphi$ *iff* $\mathfrak{M}^K, d \models \varphi$.

Proof. See the proof in the Appendix. □

5.2 Translation into neighborhood models

Readers familiar with neighborhood semantics may have already discovered that operators in $\mathcal{L}$ can be interpreted on neighborhood models in a natural way, and that **RL**-models can be translated into modally equivalent neighborhood models. Details can be found in Appendix C.

6 Conclusions and future work

This paper represents our first attempt to study the reasoning and mechanisms behind recommendation systems and in particular Content-Based Filtering models. To this end, a new logic has been proposed. We have studied its expressivity by introducing a new notion of bisimulation and translating it into a 3-variable fragment of two-sorted first-order logic. We have explored its computational properties and proved that **RL** has the tree model property, and its model-checking problem can be solved in polynomial time, for which we proposed an algorithm and proved its correctness. Finally, we also included a comparison between our models and purely Kripke or neighborhood models.

For future work, on the technical side, we want to see whether a complete axiomatization for **RL** is possible, and we are interested in exploring the complexity of its SAT problem. Inspired by the application, our models have distinct features compared to temporal models, so we want to compare them in detail. Revisiting the reasoning behind recommendations, numerous new challenges await. For instance, we would like to incorporate additional elements into our logic, allowing it to distinguish between durable goods (e.g., printers) and non-durable goods (e.g., copy paper), and reason about the periodicity of purchases, groups of agents, their interests and similarities, etc. This would connect our logical work with widely used recommendation algorithms, such as Collaborative Filtering ([15], [9], [16]).

Acknowledgements
We would like to thank Sonja Smets, Valentin Goranko, Dazhu Li, Wesley Holliday, Tomasz Klochowicz, and the four reviewers for their valuable comments on this paper. This work has been represented at the *Tsing Ch'a Session* at Tsinghua, and we thank all the audience for their inspiring suggestions.

Appendix

A Remarks on Example 1.1

The *branching structure* in Figure 1 does not represent the branching of the underlying temporal precedence relation, but is used to model varying shopping behaviors of Alice across different platforms. The arrows do not completely model the temporal precedence relation, and the extent to which they do so is in general *up to the modeler* and may be subject to *realistic restrictions*. For instance, assume that in this example, the platforms only report to a recommendation system running at d the months when certain shopping behaviors are recorded, then the recommendation system would be ignorant of the temporal precedence relation between d_2 and d_3, which are both associated with May 2023.

B Proofs

Proof. [Theorem 3.3] By induction on φ.

- The atomic and boolean cases are trivial.

- $\varphi = [R]\psi$. Assume $\mathfrak{M}, d \models \varphi$. Let $e' \in \bigcup_{i \in I'} S_i'$ such that $e' \prec' d'$. By (a) of the back condition, there exists $e \in \bigcup_{i \in I} S_i$ such that $e \prec d$ and eZe'. By the assumption, $\mathfrak{M}, e \models \psi$. By the induction hypothesis, $\mathfrak{M}', e' \models \psi$. Hence $\mathfrak{M}', d' \models \varphi$. The other direction can be proven analogously.

- $\varphi = \langle R \rangle \psi$. Assume $\mathfrak{M}, d \models \varphi$. Then there exists S_i such that

$$\text{for all } e \in S_i \text{ with } e \prec d, \text{ we have } \mathfrak{M}, e \models \varphi \tag{B.1}$$

By (b) of the forth condition, there exists S_j' such that

$$\text{for each } e' \in S_j' \text{ with } e' \prec' d', \text{ there exists } e \in S_i \text{ such that } e \prec d \text{ and } eZe' \tag{B.2}$$

Let $e' \in S_j'$ with $e' \prec' d'$. By (B.2), there exists $e \in S_i$ such that $e \prec d$ and eZe'. By (B.1), $\mathfrak{M}, e \models \psi$. By the induction hypothesis, $\mathfrak{M}', e' \models \psi$. Hence $\mathfrak{M}', d' \models \varphi$. The other direction can be proven analogously.

$\square$

Proof. [Theorem 3.5] It suffices to show that $\equiv: \mathfrak{M}, d \leftrightarrow \mathfrak{M}', d'$.

(i). The atomic condition is immediate.

(ii).(a). Suppose $\mathfrak{M}, d \equiv \mathfrak{M}', d'$, $e \in \bigcup_{i \in I} S_i$, and $e \prec d$. Assume for the sake of contradiction that there is no $e' \in \bigcup_{i \in I'} S_i'$ such that $e' \prec' d'$ and $\mathfrak{M}, e \equiv \mathfrak{M}', e'$. Since $\mathfrak{M}'$ is image-finite, $E' = \prec'^{-1}[d'] \cap \bigcup_{i \in I'} S_i'$ is finite, say $E' = \{e_1', \ldots, e_n'\}$. By the assumption, for each $e_i' \in E'$ there exists $\varphi_i \in \mathcal{L}$ such that $\mathfrak{M}, e \models \varphi_i$ but $\mathfrak{M}', e_i' \not\models \varphi_i$. Then $\mathfrak{M}, d \models \langle R \rangle \bigwedge_{1 \leq i \leq n} \varphi_i$ but $\mathfrak{M}', d' \not\models \langle R \rangle \bigwedge_{1 \leq i \leq n} \varphi_i$, contradicting $\mathfrak{M}, d \equiv \mathfrak{M}', d'$.

(ii).(b). Suppose $\mathfrak{M}, d \equiv \mathfrak{M}', d'$ and $l \in I$. Assume for the sake of contradiction that there is no S_j' such that for each $e' \in S_j'$ with $e' \prec' d'$, there exists $e \in S_l$ such that $e \prec d$ and $\mathfrak{M}, e \equiv \mathfrak{M}', e'$. Since $\mathfrak{M}'$ and $\mathfrak{M}$ are image-finite, $\{S_i'\}_{i \in I'}$ and $\prec^{-1}[d] \cap S_l$ are finite, say $\{S_i'\}_{i \in I'} = \{T_1', \ldots, T_k'\}$ and $\prec^{-1}[d] \cap S_l = \{e_1, \ldots, e_n\}$. By the assumption, for each T_i', there exists $e_i' \in T_i'$ such that $\mathfrak{M}, e_j \not\equiv \mathfrak{M}', e_i'$ for each e_j. Then for each $1 \leq i \leq k$ and

$1 \leq j \leq n$, there exists $\varphi_{ij} \in \mathcal{L}$ such that $\mathfrak{M}, e_j \models \varphi_{ij}$ but $\mathfrak{M}', e_i' \not\models \varphi_{ij}$. Let

$$\varphi = \bigvee_{1 \leq j \leq n} \bigwedge_{1 \leq i \leq k} \varphi_{ij}$$

One can show that $\mathfrak{M}, e_j \models \varphi$ for each e_j, and $\mathfrak{M}', e_i' \not\models \varphi$ for each e_i'. Hence $\mathfrak{M}, d \models \langle R \rangle \varphi$ but $\mathfrak{M}', d' \not\models \langle R \rangle \varphi$, contradicting $\mathfrak{M}, d \equiv \mathfrak{M}', d'$.

(iii). Similar to (ii). □

Proof. [Theorem 3.9] (i) By induction on φ.

- The atomic and boolean cases are trivial.

- $\varphi = [R]\psi$. We have

  $\mathfrak{M}, d \models [R]\psi$

  $\Longleftrightarrow$ for all $i \in I$ and all $d' \in S_i$ with $d' \prec d$, we have $\mathfrak{M}, d' \models \psi$

  $\Longleftrightarrow$ for all $i \in I$ and all $d' \in S_i$ with $d' \prec d$, we have $\mathfrak{M}^* \models ST_y(\psi)[d']$

  (Induction Hypothesis)

  $\Longleftrightarrow \mathfrak{M}^* \models \forall s \forall y (yEs \wedge yRx \to ST_y(\psi))[d]$

  $\Longleftrightarrow \mathfrak{M}^* \models ST_x(\varphi)[d]$

  Similarly, one can show that $\mathfrak{M}, d \models [R]\psi$ iff $\mathfrak{M}^* \models ST_y(\varphi)[d]$.

- $\varphi = \langle R \rangle \psi$. Similar to the above case.

(ii) follows directly from (i).

□

Proof. [Theorem 4.2] Assume φ is satisfied in an **RL**-model $\mathfrak{M} = (D, \{S_i\}_{i \in I}, \prec, V)$ at point $d \in D$. Let $\mathfrak{M}^d = (D^d, \{S_i^d\}_{i \in I}, \prec^d, V^d)$ be the transitive closure of the unraveling of $\mathfrak{M}$ around d. Let

$$Z = \{(e, \sigma) \in D \times D^d : (\sigma)_0 = e\}$$

By Theorem 3.3, it suffices to show that $Z : \mathfrak{M}, d \leftrightarrow \mathfrak{M}^d, (d)$.

(i) The atomic condition is immediate.

(ii).(a) Suppose $e \in \bigcup_{i \in I} S_i$ such that $e \prec d$. Then $(e, d) \in D^d$, $(e, d) \prec^d (d)$, $eZ(e, d)$, and there exists $i \in I$ such that $e \in S_i$. Then $(e, d) \in S_i^d$, so $(e, d) \in \bigcup_{i \in I} S_i^d$.

(ii).(b) Suppose $i \in I$ and $\sigma = (e_n, \ldots, e_1, d) \in S_i^d$ such that $\sigma \prec^d (d)$. Then $e_n \in S_i$, $e_n Z \sigma$, and $e_n \prec e_{n-1} \prec \cdots \prec e_1 \prec d$. By the transitivity of $\prec$, we have $e_n \prec d$.

(iii).(a) Suppose $\sigma = (e_n, \ldots, e_1, d) \in \bigcup_{i \in I} S_i^d$ such that $\sigma \prec^d (d)$. Then there exists $i \in I$ such that $\sigma \in S_i^d$, then $e_n \in S_i$. Similar to (ii).(b), it is easy to see that $e_n \prec d$.

(iii).(b) Suppose $i \in I$ and $e \in S_i$ such that $e \prec d$. Then $(e, d) \in S_i^d$, $(e, d) \prec^d (d)$, and $eZ(e, d)$. □

Proof. [Lemma 4.8](ii) is an easy consequence of (i). We show (i) by induction on φ.

- The atomic and boolean cases are trivial.
- $\varphi = \langle R \rangle \psi$. It is easy to see that after executing the loop on lines 12-14 of Algorithm 1, $S = \bigcup_{i \in I} S_i$, and after executing the loop on lines 16-18, $T = \bigcup_{d \in \text{SAT}(\mathfrak{M}, \psi) \cap S} \prec [d] = \bigcup_{d \in \text{SAT}(\mathfrak{M}, \psi) \cap \bigcup_{i \in I} S_i} \prec [d]$. By the induction hypothesis,

$$T = \bigcup_{d \in Sat^{\mathfrak{M}}(\psi) \cap \bigcup_{i \in I} S_i} \prec [d]$$

Obviously $\text{SAT}(\mathfrak{M}, \varphi) = T = Sat^{\mathfrak{M}}(\varphi)$.

- $\varphi = [R]\psi$. It is easy to see that after executing the inner loop on lines 25-27, $E = \bigcup_{d \in \text{SAT}(\mathfrak{M}, \psi) \cap S_i} \prec [d]$, so after executing the outer loop on lines 23-29, $T = \bigcap_{i \in I} \bigcup_{d \in \text{SAT}(\mathfrak{M}, \psi) \cap S_i} \prec [d]$. By the induction hypothesis,

$$T = \bigcap_{i \in I} \bigcup_{d \in Sat^{\mathfrak{M}}(\psi) \cap S_i} \prec [d]$$

Obviously $\text{SAT}(\mathfrak{M}, \varphi) = T = Sat^{\mathfrak{M}}(\varphi)$.

$\square$

Proof. [Theorem 4.6] By Lemma 4.8, it suffices to show that the time complexity of the function CHECK in Algorithm 1 is in $O(|I| \cdot (|D| + | \prec |) \cdot |\varphi|)$. Let $\mathfrak{M} = (D, \{S_i\}_{i \in I}, \prec, V)$ be a finite **RL**-model, $d \in D$, and $\varphi \in \mathcal{L}$.

Claim. For each $\varphi \in \mathcal{L}$, if $\text{SAT}(\mathfrak{M}, \psi)$ has been computed for each subformula ψ of φ, then $\text{SAT}(\mathfrak{M}, \varphi)$ can be done in $O(|I| \cdot (|D| + | \prec |))$.

Proof of the Claim. When $\varphi = \bot$, $\text{SAT}(\mathfrak{M}, \varphi)$ can finish in $O(1)$. When $\varphi \in \textbf{Prop}$ or $\varphi = \neg \psi$ or $\varphi = \psi_1 \wedge \psi_2$, $\text{SAT}(\mathfrak{M}, \varphi)$ can finish in $O(|D|)$.

Suppose $\varphi = \langle R \rangle \psi$. The first loop on lines 12-14 can be done in $O(|I| \cdot |D|)$. The second loop on lines 16-18 first requires $O(|D|)$ time to compute $\text{SAT}(\mathfrak{M}, \psi) \cap S$, then visits each edge in $\prec$ at most once, which requires $O(| \prec |)$ time. Therefore $\text{SAT}(\mathfrak{M}, \varphi)$ can be done in $O(|I| \cdot |D| + | \prec |)$.

Suppose $\varphi = [R]\psi$. Line 22 can be done in $O(|D|)$. The inner loop on lines 25-27 first requires $O(|D|)$ time to compute $G \cap S_i$, then visits each edge in $\prec$ at most once, which requires $O(| \prec |)$ time. Line 28 takes $O(|D|)$ time. Therefore both the outer loop and $\text{SAT}(\mathfrak{M}, \varphi)$ can be done in $O(|I| \cdot (|D| + | \prec |))$.

Hence $\text{SAT}(\mathfrak{M}, \varphi)$ can be done in $O(|I| \cdot (|D| + | \prec |))$.

$\square$(of the Claim)

Note that SAT is a bottom-up traversal of the parse tree of φ, starting from the leaves and finishing at the root. By the Claim, visiting each node of the parse tree of φ takes $O(|I| \cdot (|D| + | \prec |))$ time, and there are at most $|\varphi|$ such nodes, so the time complexity of SAT and thus CHECK is in $O(|I| \cdot (|D| + | \prec |) \cdot |\varphi|)$.

$\square$

Proof. [Fact 5.3] Irreflexivity of $\bigcup_{i \in I \cup \{I\}} \prec_i$: Let $d \in D$ and $i \in I \cup \{I\}$, then $d \not\prec d$ by the irreflexivity of $\prec$, then $d \not\prec_i d$, so $(d, d) \notin \bigcup_{i \in I \cup \{I\}} \prec_i$.

Transitivity of $\bigcup_{i \in I \cup \{I\}} \prec_i$: Suppose $(d_1, d_2), (d_2, d_3) \in \bigcup_{i \in I \cup \{I\}} \prec_i$. Then there exist $i, j \in I \cup \{I\}$ such that $d_1 \prec_i d_2$ and $d_2 \prec_j d_3$. Then $d_1 \prec d_2$ and $d_2 \prec d_3$. By the transitivity of $\prec$, $d_1 \prec d_3$, so $d_1 \prec_i d_3$. Hence $(d_1, d_3) \in \bigcup_{i \in I \cup \{I\}} \prec_i$.

Forward consistency: Suppose $\prec_i [d] \neq \emptyset$ and $\prec_j [d] \neq \emptyset$. It cannot be the case where exactly one of i of j is I, for that in that case we would have both $d \in \bigcup_{i \in I} S_i$ and $d \notin \bigcup_{i \in I} S_i$, which is not possible. Then there are two cases:

(i) $i = j$. Obviously $\prec_i [d] = \prec_j [d]$.

(ii) $i \neq j$, and neither i nor j is I. Then $d \in S_i$ and $d \in S_j$. So $d' \in \prec_i [d]$ iff $d \prec d'$ iff $d' \in \prec_j [d]$.

$\square$

Proof. [Theorem 5.6]
Claim. For each $d, d' \in D$ and $i \in I$, we have $d' \prec_i d$ iff $d' \in S_i$ and $d' \prec d$.
Proof of the Claim. Suppose $d' \in S_i$ and $d' \prec d$. Assume $d' \not\prec_i d$. Then there exists $j \in I \cup \{I\}$ such that $j \neq i$ and $d' \prec_j d$. Since $\prec_i [d'] \neq \emptyset$ and $\prec_j [d'] \neq \emptyset$, by the forward consistency of $\mathfrak{K}$, we have $\prec_i [d'] = \prec_j [d']$, so $d' \prec_i d$, contradicting $d' \not\prec_i d$. Hence $d' \prec_i d$.

The other direction is straightforward.

$\square$(of the Claim)

By induction on φ.

- The atomic and boolean cases are trivial.

- $\varphi = [R]\psi$. We have

$\mathfrak{K}, d \models \varphi$
$\Longleftrightarrow$ for all $i \in I$ and all $d' \in D$ with $d' \prec_i d$, we have $\mathfrak{K}, d' \models \psi$
$\Longleftrightarrow$ for all $i \in I$ and all $d' \in D$ with $d' \prec_i d$, we have $\mathfrak{M}^K, d' \models \psi$
  (Induction Hypothesis)
$\Longleftrightarrow$ for all $i \in I$ and all $d' \in S_i$ with $d' \prec d$, we have $\mathfrak{M}^K, d' \models \psi$
  (Claim)
$\Longleftrightarrow \mathfrak{M}^K, d \models \varphi$

- $\varphi = \langle R]\psi$. We have

$\mathfrak{K}, d \models \psi$
$\Longleftrightarrow$ there exists $i \in I$ such that for all $d' \in D$ with $d' \prec_i d$, $\mathfrak{K}, d' \models \psi$
$\Longleftrightarrow$ there exists $i \in I$ such that for all $d' \in D$ with $d' \prec_i d$, $\mathfrak{M}^K, d' \models \psi$
  (Induction Hypothesis)
$\Longleftrightarrow$ there exists $i \in I$ such that for all $d' \in S_i$ with $d' \prec d$, $\mathfrak{M}^K, d' \models \psi$
  (Claim)
$\Longleftrightarrow \mathfrak{M}^K, d \models \varphi$

$\square$

C Neighborhood RL-models

We first define the neighborhood counterpart of **RL**-models and truth on those neighborhood models.

Definition C.1 [Neighborhood **RL**-model] A *neighborhood* **RL**-*model* is defined as a triple (D, N, V), where

(i) $N : D \to \mathcal{P}(\mathcal{P}(D))$ is a *neighborhood function*.

(ii) $V : \mathbf{Prop} \to \mathcal{P}(D)$ is a *valuation*.

Definition C.2 [Truth] The *truth* of an arbitrary formula $\varphi \in \mathcal{L}$ in a neighborhood **RL**-model $\mathfrak{N} = (D, N, V)$ at $d \in D$ is defined inductively as follows:

$$
\begin{aligned}
\mathfrak{N}, d \models p & \quad \text{iff} \quad d \in V(p) \\
\mathfrak{N}, d \models \bot & \quad \text{iff} \quad \text{never} \\
\mathfrak{N}, d \models \neg\varphi & \quad \text{iff} \quad \mathfrak{N}, d \nvDash \varphi \\
\mathfrak{N}, d \models \varphi \wedge \psi & \quad \text{iff} \quad \mathfrak{N}, d \models \varphi \text{ and } \mathfrak{N}, d \models \psi \\
\mathfrak{N}, d \models [R]\varphi & \quad \text{iff} \quad \text{for all } X \in N(d) \text{ and all } d' \in X, \text{ we have } \mathfrak{N}, d' \models \varphi \\
\mathfrak{N}, d \models \langle R \rangle \varphi & \quad \text{iff} \quad \text{there exists } X \in N(d) \text{ such that for all } d' \in X, \text{ we} \\
& \qquad\qquad \text{have } \mathfrak{N}, d' \models \varphi
\end{aligned}
$$

We then show how to translate an **RL**-model into a neighborhood **RL**-model without affecting truth of $\mathcal{L}$-formulas.

Definition C.3 [Neighborhood translation of **RL**-models] Given an **RL**-model $\mathfrak{M} = (D, \{S_i\}_{i \in I}, \prec, V)$, the *neighborhood translation of* $\mathfrak{M}$ is a neighborhood **RL**-model $\mathfrak{N}^M = (D, N, V)$, where $N(d) = \{S_i \cap \prec^{-1}[d] : i \in I\}$ for each $d \in D$.

Theorem C.4 *Let* $\mathfrak{M} = (D, \{S_i\}_{i \in I}, \prec, V)$ *be an* **RL**-*model, and* $\mathfrak{N}^M$ *be the neighborhood translation of* $\mathfrak{M}$. *For each* $d \in D$, $\mathfrak{M}, d \equiv \mathfrak{N}^M, d$.

Proof. By induction on φ.

- The atomic and boolean cases are trivial.

- $\varphi = [R]\psi$. We have

$$
\begin{aligned}
& \mathfrak{M}, d \models \varphi \\
\iff & \text{for all } i \in I \text{ and all } d' \in S_i \cap \prec^{-1}[d], \text{ we have } \mathfrak{M}, d' \models \psi \\
\iff & \text{for all } i \in I \text{ and all } d' \in S_i \cap \prec^{-1}[d], \text{ we have } \mathfrak{N}^M, d' \models \psi \\
& (\text{Induction Hypothesis}) \\
\iff & \text{for all } X \in N(d) \text{ and all } d' \in X, \text{ we have } \mathfrak{N}^M, d' \models \psi \\
\iff & \mathfrak{N}^M, d \models \varphi
\end{aligned}
$$

- $\varphi = \langle R \rangle \psi$. We have

$$
\begin{aligned}
& \mathfrak{M}, d \models \varphi \\
\iff & \text{there is } i \in I \text{ such that for all } d' \in S_i \cap \prec^{-1}[d], \text{ we have } \mathfrak{M}, d' \models \psi
\end{aligned}
$$

$\Longleftrightarrow$ there is $i \in I$ such that for all $d' \in S_i \cap \prec^{-1} [d]$, we have $\mathfrak{N}^M, d' \models \psi$ (Induction Hypothesis)

$\Longleftrightarrow$ there is $X \in N(d)$ such that for all $d' \in X$, we have $\mathfrak{N}^M, d' \models \psi$

$\Longleftrightarrow \mathfrak{N}^M, d \models \varphi$

$\square$

References

[1] Baier, C. and J.-P. Katoen, "Principles of Model Checking," The MIT Press, Cambridge, MA, 2008.

[2] Belardinelli, G., L. Li, S. Smets and A. Solaki, *Logics for personalized announcements and attention dynamics*, Manuscript (2024).

[3] Blackburn, P., M. de Rijke and Y. Venema, "Modal Logic," Cambridge Tracts in Theoretical Computer Science, Cambridge University Press, New York, 2001.

[4] Cena, F., L. Console and F. Vernero, *Logical foundations of knowledge-based recommender systems: A unifying spectrum of alternatives*, Information Sciences **546** (2021), pp. 60–73.

[5] Farahbakhsh, R., A. Cuevas and N. Crespi, *Characterization of cross-posting activity for professional users across Facebook, Twitter and Google+*, Social Network Analysis and Mining **6** (2016), pp. 1–14.

[6] Han, J., M. Kamber and J. Pei, "Data Mining: Concepts and Techniques," The Morgan Kaufmann Series in Data Management Systems, Morgan Kaufmann, 2012, 3rd edition.

[7] Heimbach, I., B. Schiller, T. Strufe and O. Hinz, *Content virality on online social networks: Empirical evidence from Twitter, Facebook, and Google+ on German news websites*, in: *Proceedings of the 26th ACM Conference on Hypertext & Social Media*, HT '15 (2015), pp. 39–47.

[8] Jain, A. and C. Gupta, *Fuzzy logic in recommender systems*, in: O. Castillo, P. Melin and J. Kacprzyk, editors, *Fuzzy Logic Augmentation of Neural and Optimization Algorithms: Theoretical Aspects and Real Applications*, Studies in Computational Intelligence **749**, Springer International Publishing AG, Cham, 2018 pp. 255–274.

[9] Ko, H., S. Lee, Y. Park and A. Choi, *A survey of recommendation systems: Recommendation models, techniques, and application fields*, Electronics **11** (2022).

[10] Lee, R. K.-W., T.-A. Hoang and E.-P. Lim, *On analyzing user topic-specific platform preferences across multiple social media sites*, in: *Proceedings of the 26th International Conference on World Wide Web*, WWW '17 (2017), pp. 1351–1359.

[11] Leskovec, J., A. Rajaraman and J. D. Ullman, "Mining of Massive Datasets," Cambridge University Press, 2014, 2nd edition.

[12] Li, L., "Games, Boards and Play: A Logical Perspective," Ph.D. Thesis, University of Amsterdam, 2023.

[13] Pacuit, E., "Neighborhood Semantics for Modal Logic," Short Textbooks in Logic, Springer International Publishing AG, Cham, 2017.

[14] Patro, S. G. K., B. K. Mishra, S. K. Panda and R. Kumar, *Fuzzy logics based recommendation systems in e-commerce: A review*, in: P. K. Pattnaik, M. Sain and A. A. Al-Absi, editors, *Proceedings of 2nd International Conference on Smart Computing and Cyber Security* (2022), pp. 107–120.

[15] Ricci, F., L. Rokach, B. Shapira and P. B. Kantor, "Recommender Systems Handbook," Springer, 2011.

[16] Roy, D. and M. Dutta, *A systematic review and research perspective on recommender systems*, Journal of Big Data **9** (2022).

[17] Samuelson, P. A., *A note on the pure theory of consumer's behaviour*, Economica **5** (1938), pp. 61–71.

[18] van Benthem, J., "Modal Logic for Open Minds," CSLI Publications, Stanford, CA, 2010.

[19] Varian, H. R., *The nonparametric approach to demand analysis*, Econometrica **50** (1982), pp. 945–973.

[20] Xue, Y., C. Xiao, X. Luo and W. Yang, *Predicting platform preference of online contents across social media networks*, IEEE Access **7** (2019), pp. 136428–136438.

Enforce Actions Based on Structured Argumentation Theory under Legal Contexts

Yiwei Lu

School of Law, Old College, University of Edinburgh
Edinburgh, UK
Y.Lu-104@sms.ed.ac.uk

Zhe Yu [1]

Institute of Logic and Cognition, Department of Philosophy, Sun Yat-sen University
Guangzhou, China
zheyusep@foxmail.com

Abstract

In this paper, we explore the necessary reasoning conditions for outputting certain targeted actions in a structured argumentation system in the context of law. Based on a system that has been developed to combine legal ontology with a structured argumentation framework, we adapt the norms in the law regarding the actions of AI products by means of this feature. The adjustment of norms is realised under non-monotonic reasoning through argumentation theory, and dynamic changes in preference order among legal principles are also reflected.

Keywords: Enforcement, Structured argumentation, Practical reasoning.

1 Introduction

Structured argumentation has shown powerful ability in non-monotonic reasoning and dealing with conflicts over the last few years. How to explore the interpretability of reasoning results in different inconsistent contexts has been a direction of great concern in recent years [10,12]. Among them, research incorporating structured argumentation has become increasingly well-studied. These studies focus on how to get results from inconsistent reasoning conditions and have many combinations with practical reasoning, e.g., decision-making. Reasoning, however, is not limited to the process of moving from conditions to outcomes. Analysing the corresponding reasoning structures through a given goal or outcome is also an important form of reasoning in life. For example, the formulation of social rules embedding complex principles of reasoning often

---

[1] Corresponding author. This research is supported by the National Social Science Foundation of China (No.20&ZD047).

starts with a predetermined goal. It is, therefore, natural to notice the opposite direction: how can the results of reasoning be used to determine the necessary inferential conditions for arriving at it?

This is particularly evident in practical reasoning, especially legal reasoning. If the law is understood as a set of social rules providing basic guidance for the behavior of citizens, the first step in the formulation of these rules is a clear legislative intention. At a macro level, the intention of the law may be human rights or philosophical concepts such as freedom and equality. The same is true for more specific legal norms. For example, norms about direction of travel, speed, and driving style are established because the law wants vehicles to be driven without collision. It can be seen that legal norms are developed on the basis of common-sense rules, such as the rules of physics. However, the connections between legal norms and the principles of reasoning are designed around legislative intention. Building reasoning rules around legislative intention is likewise one of the reasons why legal norms need to be constantly adapted. One such situation is when existing legal rules no longer fulfill the same legal intention in new situations. For example, the law prohibits drivers from using communication devices or screen devices while driving because it distracts the driver's attention from the road conditions. In the case of autonomous driving, however, the use of communication and visualisation devices is exactly how the driving system gets the road conditions.

In the other case, which is the main focus of this paper, the intention itself is changed. In order not to cause confusion, it should be made clear at the outset that the intention in this paper is only with respect to the specific action that the rule seeks to ensure, such as ensuring that a vehicle stops in front of people and not with respect to abstract concepts such as safety or justice. Even at a level very close to the application, the update of the law's guidance on actions is not simply a matter of modifying the ultimate direction of a rule but a more profound renewal of the relationships between the rules and the corresponding legal principles in light of the new intentions. This is not just because the legal system is so complex that a single change can have a knock-on effect, so more work is required to ensure self-consistency. It is also because Dworkin's theory of the "one right answer" [21] rarely occurs in practical reasoning in the field of law. For the most part, the application of the legal rules is subject to a process of interpretation, i.e., the selection of the applicable rules on the basis of conflicting or uncertain elements of the case and, more importantly, the priority of the legal principles. Therefore, the update of rules of legal reasoning according to the new intentions must also capture the characteristics of such non-monotonic reasoning and reflect changes in the preferences of legal principles.

We use the following example to exemplify the above notions and will also use it later in the paper to show how the findings can be useful in similar contexts.

Example 1.1 *Under existing traffic regulations, when people and goods are involved in a traffic accident or other dangerous situation at the same time, the* *driver's life will be protected as a priority. This is, in fact, underpinned by a legal principle that contains a message of preference:*

Driver safety has priority over cargo safety

The action that the above legal principle seeks to ensure is that the safety of the driver must be protected in the event of danger. However, the advent of autonomous driving may change this intention. This is because in an autonomous driving situation, the driver is no longer a human being but some system or the car itself, which may not be as important as their cargo. In a more extreme scenario, we could envisage a self-driving car with a dog in it - who should we protect in danger? Based on the legal principle above, the car would make the decision to sacrifice the dog for its own safety. This is because it fits the notion of a driver, and in the legal context, pets or livestock are property, a type of cargo. But it is obvious that there will be a significant number of people who do not want the car to make this choice because it seems that at this moment, the legal principle that should be applied instead of the previous one is:

Lives have priority over inanimate objects

Therefore, in this case, the action that we have to ensure has changed to the safety of lives, or important cargoes must be protected. In this case, it is the safety of the dog must be protected. Accordingly, the rules of law need to be modified.

As we can see from this example, the revision of legal rules in a new situation is a process in which new elements interact with the intention of the law. The process begins with the ability to express information about the new environment; in this case, the changes brought about by the new design elements, such as the driver, do not fit in the human concept anymore. This is followed by a change in the action that the legal reasoning seeks to secure in light of the corresponding change. Moreover, in order to realise this new legal intent, the applicability of the legal principles in that situation is determined by a reordering of preferences. Through this, it is ultimately determined what the new rules and the relationship between them should be. In this case, perhaps one option would be to stipulate that in the case of autonomous driving, the action to protect the dog in the event of danger always defeats the action to protect the driver should be enforced.

In order to explore how to use structured argumentation systems to help implement the above reasoning in practical reasoning of law, we have chosen to do so within a framework called *LeSAC* (legal support system for autonomous cars) [16,17,15]. Since it is used only as a theoretical background for this study, we will not give a detailed description of it here but rather a brief introduction, and will provide some of the basic definitions needed later on. According to Figure 1, *LeSAC* is a structured argumentative framework based on the legal ontology. It extends the ability of ontology technology to create a common vocabulary with the storage of legal information through structured argumentation, giving it the ability to reason non-monotonically. It can describe legal information with the ability to deal with conflicts between law and design and supports reasoning under preferences. This has been designed to be used in previous research to enable legal compliance detection and modification options for AI product design. In contrast, this paper hopes to help achieve the aim of updating the conditions of legal reasoning through the new legal intentions brought about by the features of AI products via the functionality of structured argumentation.

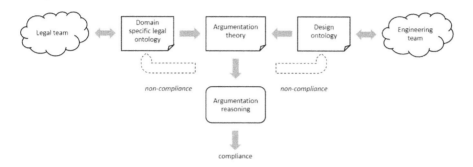

Fig. 1. Overview of the process of *LeSAC*

This paper is organized in the following manner. Section 2 is devoted to examining related work. The following section first introduces the basic settings of the structured argumentation theory *LeSAC* , then we discuss how to determine conditions under which a certain action can be considered sceptically and credulously justified and how the law could be adjusted by so, demonstrating with Example 1.1. In section 4, we summarize this work and outline directions for future research.

2 Related Works

In fact, the repair of the reasoning framework given a reasoning outcome is a constant concern [5]. A methodology for modifying faulty logical theories in the framework of classical logic with respect to a targeted outcome [8] has been developed as a logical framework repair tool called ABC [9] and applied to the modification of legal rules [13]. The study is able to give a rich variety of repair options given the conclusions that must be secured, such as creating new concepts, adding or deleting rules of inference, etc. However, this and other modification methods based on classical logical approaches are not able to support non-monotonic reasoning. And they are more focused on resolving inconsistencies in formal logic and thus pursuing self-consistency, missing the realistic correspondence and explainability of the reasoning process. Therefore, we still believe that argumentation theory has its unique advantages in this problem.

Since the development of formal argumentation, the enforcement of arguments has attracted a lot of attention from researchers over the decades. Most existing literature focuses on abstract rather than structured argumentation frameworks [3,2,22]. These works summarize the methods and principles of how to adjust the abstract argumentation framework to make an argument acceptable, as well as how to choose efficient methods, etc. However, in terms of legal applications, it is challenging for them to reflect the defensive process required by law, as they do not involve the internal structure of arguments. Currently, there are relatively few studies addressing this issue from the perspective of structured argumentation [4,20]. While these studies place less emphasis on specific application contexts, they have not explicitly discussed the role of argumentation preferences in this process. This paper therefore chooses to build on *LeSAC* in an attempt to achieve an update of the conditions for preference-inclusive reasoning in a specific legal context.

3 Enforcement of Actions

3.1 An argumentation system for autonomous vehicle design

As mentioned above, in [16,17,15], we built a legal support argumentation theory for autonomous vehicle designs based on legal ontology and structured argumentation systems [6,19,18], namely *LeSAC* . Here we give some basic concepts and definitions to help understand the following content.

Let $\Delta = (T, A)$ be a legal ontology for autonomous vehicle design based on Description Logics (DL), where T and A represent the *TBox*, which introduces the terminology, and *ABox*, which contains facts about individuals of an ontology, respectively. Given an argumentation system AS, $(AS, \mathcal{K}^A)$ is an argumentation theory about Δ, where $AS = (\mathcal{L}, \mathcal{R}^T, n)$ such that $\mathcal{R}^T$ is the set of rules corresponding to T (a mapping table can be found in [14]), and $\mathcal{K}^A$ is the set of premises based on A. Formally, a *LeSAC* can be defined as follows.

Definition 3.1 [LeSAC] Let $\Delta = (T, A)$ be a legal ontology, $LeSAC = (\mathcal{L}, \mathcal{R}^T, n, \mathcal{K}^A)$ is an argumentation theory instantiated by Δ, where:

- $\mathcal{L}$ is a formal language based on description logic and closed under negation ($\neg$), where $\psi = -\varphi$ denotes $\psi = \neg\varphi$ or $\varphi = \neg\psi$.
- $\mathcal{R}^T = \mathcal{R}_s \cup \mathcal{N}$ is a set of rules corresponding to T, where $\mathcal{R}_s$ is a set of strict inference rules of the form $\varphi_1, \ldots, \varphi_n \to \varphi$, and $\mathcal{N}$ is a set of legal norms of the form $\varphi_1, \ldots, \varphi_n \Rightarrow \varphi$ ($\varphi_i, \varphi \in \mathcal{L}$); let $\mathcal{R}_s \cap \mathcal{N} = \emptyset$.
- n is a naming function such that $n : \mathcal{N} \to \mathcal{L}$.
- $\mathcal{K}^A$ is a knowledge base based on A.

Let $\mathtt{Prem}(\alpha)$ returns the set of all the formulas of $\mathcal{K}$ used to build argument α, $\mathtt{Conc}(\alpha)$ returns the conclusion of α, $\mathtt{Sub}(\alpha)$ returns the set of all the subarguments of α, and $\mathtt{Rules}(\alpha)$ returns the set of all rules applied in α. Arguments are constructed by rules from the knowledge base, defined as follows.

Definition 3.2 [Argument] An argument α based on a *LeSAC* is a structure obtained by applying one or more of the following steps finitely many times:

   (i) φ, if $\varphi \in \mathcal{K}^A$, s.t. $\mathtt{Prem}(\alpha) = \{\varphi\}$, $\mathtt{Conc}(\alpha) = \varphi$ and $\mathtt{Rules}(\alpha) = \emptyset$;

(ii) $\alpha_1, \ldots, \alpha_n \dashrightarrow \psi$,[2] if $\alpha_1, \ldots, \alpha_n$ are arguments, such that there exists a rule $\mathtt{Conc}(\alpha_1), \ldots, \mathtt{Conc}(\alpha_n) \dashrightarrow \psi$ in $\mathcal{R}^T$, and $\mathtt{Prem}(\alpha) = \mathtt{Prem}(\alpha_1) \cup \ldots \cup \mathtt{Prem}(\alpha_n)$, $\mathtt{Conc}(\alpha) = \psi$, $\mathtt{Sub}(\alpha) = \mathtt{Sub}(\alpha_1) \cup \ldots \cup \mathtt{Sub}(\alpha_n) \cup \{\alpha\}$, $\mathtt{Rules}(\alpha) = \mathtt{Rules}(\alpha_1) \cup \ldots \cup \mathtt{Rules}(\alpha_n) \cup \{\mathtt{Conc}(\alpha_1), \ldots, \mathtt{Conc}(\alpha_n) \dashrightarrow \psi\}$.

Conflicts between arguments are called attacks, while the defeat relation is determined by the preferences on the set of arguments, defined as follows.

Definition 3.3 [Attacks and defeats] Let α, β, β' be arguments constructed based on a *LeSAC* , and $\preceq$ a preference ordering on $\mathcal{A}$. α **attacks** β on β', iff $\mathtt{Conc}(\alpha) = -\varphi$ and: 1) $\beta' \in \mathtt{Sub}(\beta)$ of the form $\beta_1'', \ldots, \beta_n'' \Rightarrow \varphi$; or 2) $\beta' = \varphi$ and $\varphi \in \mathtt{Prem}(\beta) \cap \mathcal{K}^A$.

Then α **defeats** β, iff α attacks β on β' and $\alpha \nprec \beta'$. [3]

We say that α strictly defeats β, if α defeats β, while β does not defeat α.

Based on the set of arguments and the defeat relation between arguments, abstract argumentation frameworks can be constructed and arguments are evaluated based on argumentation semantics. The following argumentation semantics are defined according to [7].

Definition 3.4 [Argumentation semantics] Let $\mathcal{A}$ be the set of all the constructed arguments based on a *LeSAC* , $att = \mathcal{A} \times \mathcal{A}$ the set of attacks, and $\mathcal{D} \subseteq att$ is a set of defeats. An abstract framework AF is a tuple $(\mathcal{A}, \mathcal{D})$. We say an extension $E \subseteq \mathcal{A}$ is *conflict-free* iff $\nexists \alpha, \beta \in E$ s.t. $(\alpha, \beta) \in \mathcal{D}$, and α is *defended* by E (or *acceptable* w.r.t. E), iff $\forall \beta \in \mathcal{A}$, if $(\beta, \alpha) \in \mathcal{D}$, then $\exists \gamma \in E$ such that $(\gamma, \beta) \in \mathcal{D}$. Then:

- E is an *admissible set* iff E is conflict-free and $\forall \alpha \in E$, α is defended by E;

- E is a *complete* extension iff E is admissible, and $\forall \alpha \in \mathcal{A}$ defended by E, $\alpha \in E$;

- E is a *grounded* extension iff E is a minimal complete extension w.r.t. set-inclusion;

- E is a *preferred* extension iff E is a maximal complete extension w.r.t. set-inclusion.

An argument α is said to be accepted/justified by an extension $E \subseteq \mathcal{A}$ under certain argumentation semantics, if and only if $\alpha \in E$. And for any set of accepted arguments $E \subseteq \mathcal{A}$, $\mathtt{Conc}(E) = \{\mathtt{Conc}(\alpha) | \alpha \in E\}$ is the corresponding set of accepted conclusions of E.

With these basic definitions, we can briefly formalize Example 1.1 in original *LeSAC* as follows:

Example 3.5 [Original *LeSAC*]

$$\mathcal{N} = \left\{ \begin{array}{l} r_1 : Driver(x), EncounterAccident(x) \Rightarrow Protect(x); \\ r_2 : Cargo(x), EncounterAccident(x) \Rightarrow \neg Protect(x) \end{array} \right\}$$

---

[2] "$\dashrightarrow$" denotes "$\rightarrow / \Rightarrow$"

[3] For any arguments α and β, $\alpha \prec \beta$ if and only if $\alpha \preceq \beta$ and $\beta \npreceq \alpha$.

$$\mathcal{K}^A = \left\{ \begin{array}{l} Driver(AV); Cargo(Dogs); \\ EncounterAccident(AV); EncounterAccident(Dogs) \end{array} \right\}$$

$$\mathcal{P} = \left\{ \begin{array}{l} p_1 : Safety\ of\ the\ driver\ should\ take\ precedence\ over\ safety\ of\ the\ cargo. \\ p_2 : Protecting\ lives\ should\ take\ precedence\ over\ protecting\ inanimate\ obj \end{array} \right.$$

In Example 3.5, $\mathcal{P}$ is a set of legal principles. In [16], we suggested that preferences on the set of arguments can be "lifted" from the priorities of legal principles that primarily underpin a rule or norm r applied in arguments, denoted as $prin(r)$. Then for Example 3.5, $prin(r_1) = p_1$, $prin(r_2) = p_1$. In this case, $LeSAC$ can describe the process by which existing regulations do not apply to new situations through argumentation reasoning. Based on previous research [16,17], it can test the legal compliance of AI product solutions provided by designers either as is or given new reasoning criteria, e.g., by telling the designer whether it is legal to design a vehicle to protect itself. And it can suggest changes to the proposal with ongoing evaluation. However, it does not support adjusting reasoning conditions to ensure that a particular behaviour is protected. We will therefore implement this feature in the next section under structured argumentation.

3.2   Enforcement of actions

Particularly in practical reasoning, actions constitute the decisions that agents need to make during the reasoning process. For instance, a system like $LeSAC$ is designed to support autonomous vehicle design in making decisions regarding actions, based on a specified legal ontology. Consequently, the output of accepted conclusions manifests as a set of actions. The primary focus of this paper is on how to ensure a certain action is justified within the reasoning system.

Based on the argumentation semantics introduced in Definition 3.4, since the preferred extensions and grounded extensions are all complete extensions, we will focus on the discussion based on the complete semantics. The sets of extensions based on these semantics are denoted as Pr, Gr, and Co, respectively.

We subsequently explore sceptical and credulous justifications separately.

3.2.1   Sceptical justification

According to the extension-based argumentation semantics, an argument is said to be sceptically accepted/justified if and only if it is an element of each extension obtained based on certain argumentation semantics [1]. Correspondingly, we say that an action a is sceptically justified if and only if there exists a sceptically accepted argument α, such that $\texttt{Conc}(\alpha) = a$.

If an argument α is sceptically accepted, then based on Definition 3.4, α is included in the unique grounded extension obtained by the same abstract argumentation framework.

This implies that there is a reasoning chain that leads to the conclusion a, or in other words, there are formulas in $\mathcal{K}^A$ and rules in $\mathcal{R}^T$ of a *LeSAC* that constitute subarguments of α, such that $\text{Conc}(\alpha) = a$, and meanwhile, for $E \in Gr$ (denoted as E_{Gr}), $\text{Sub}(\alpha) \subseteq E$. Therefore, we have the following proposition.

Proposition 3.6 *Let* $LeSAC = (\mathcal{L}, \mathcal{R}^T, n, \mathcal{K}^A)$ *be an argumentation theory,* $\mathcal{A}$ *all the arguments constructed by LeSAC , and* α *an argument. An action a is sceptically justified under complete semantics, iff*

(i) $\exists a_1, a_2, \ldots, a_j, \ldots, a_k, \ldots a_n \in \mathcal{K}^A$ *and rules* $r_1, \ldots, r_n \in \mathcal{R}^T$ *such that* $r_1 = a_1, a_2, \ldots, a_j \dashrightarrow a_1'$, $r_2 = a_1', \ldots, a_k \dashrightarrow a_2'$, $\ldots$, *and* $r_n = a_1', a_2', \ldots, a_n \dashrightarrow a$;

(ii) $\forall \alpha' \in \text{Sub}(\alpha)$, $\nexists \alpha'' \in \text{Sub}(\alpha)$ *such that* α'' *attacks* α';

(iii) $\forall \alpha' \in \text{Sub}(\alpha)$, *if* $\exists \beta \in \mathcal{A}$ *such that* β *attacks* α' *and* $\beta \nprec \alpha'$, *then* $\exists \gamma \in E_{Gr}$ *such that* $\gamma \neq \alpha'$ *and attacks* β *on* $\beta' \in \text{Sub}(\beta)$, *and* $\beta' \prec \gamma$.

Proofs of propositions can be found in the Appendix.

In Proposition 3.6, the first condition states that there is a path starting from $\mathcal{K}^A$, connected by rules in $\mathcal{R}^T$, forming an argument with the conclusion a. Based on these formulas and rules, all arguments in $\text{Sub}(A)$ can be constructed. This suggests that in order to make a certain action justified by an argumentation theory, we should supplement the premises in the knowledge base and rules in the set of rules, from which the action can be derived. Particularly, for an argumentation theory like *LeSAC* built on the basis of a legal ontology, elements in the *TBox* and *ABox* should be added. Arguments and subarguments supporting this conclusion should thus be constructed, leading to a corresponding expansion of the original abstract argumentation frameworks.

Consider Example 3.5: if we hope the dogs in the autonomous vehicle (AV) to be protected, then we want the assertion of action "*Protect(Dogs)*" to be justified. The original *LeSAC* can be updated as in Example 3.7, with a new assertion "*Life(Dogs)*" and a new rule "*Life(x), EncounterAccident(x)* $\Rightarrow$ *Protect(x)*" be added.

Example 3.7 [Updated *LeSAC*]

$$\mathcal{N} = \left\{ \begin{array}{l} r_1 : Driver(x), EncounterAccident(x) \Rightarrow Protect(x); \\ r_2 : Cargo(x), EncounterAccident(x) \Rightarrow \neg Protect(x); \\ r_3 : Life(x), EncounterAccident(x) \Rightarrow Protect(x); \\ r_4 : Inanimate(x), EncounterAccident(x) \Rightarrow \neg Protect(x) \end{array} \right\}$$

$$\mathcal{K}^A = \left\{ \begin{array}{l} Driver(AV); Cargo(Dogs); Life(Dogs); Inanimate(AV); \\ EncounterAccident(AV); EncounterAccident(Dogs) \end{array} \right\}$$

$$\mathcal{P} = \left\{ \begin{array}{l} p_1 : Safety\ of\ the\ driver\ should\ take\ precedence\ over\ safety\ of\ the\ cargo. \\ p_2 : Protecting\ lives\ should\ take\ precedence\ over\ protecting\ inanimate\ objects. \end{array} \right.$$

$$prin(r_1) = p_1; \ prin(r_2) = p_1; \ prin(r_3) = p_2; \ prin(r_4) = p_2$$

In addition, the second condition states that $\mathtt{Sub}(\alpha)$ should be conflict-free according to Definition 3.3, and the third condition states that all the arguments leading to the conclusion a are either not defeated by any argument, or are defended by an argument already included in a grounded extension through strict defeat. In many practical cases, this means that any argument in the set $\mathtt{Sub}(\alpha)$ is not attacked or is defended by arguments that are not attacked and, therefore, have to be accepted. To prevent the subarguments from being attacked, the involved attacks can be deleted by changing the preferences between arguments. At a deeper level, and based on our previous work [16], preferences on the set of arguments can be lifted from the priority relation between legal principles, suggesting designers consider whether it is necessary or possible to reasonably adjust the relevant priority orderings. Another consideration is that in some contexts, we can consider deleting those arguments that cause attacks/defeats. Last but not least, in order to defend the attacked arguments in the set of subarguments, users could consider adding arguments (that have to be accepted) and adding the corresponding attacks.

Consider the updated $LeSAC$ in Example 3.7. If we denote the argument constructed based on r_2 as α and the argument constructed based on r_3 as β, then if the principle p_1 strictly takes precedence over p_2, after preference lifting, we might get $\alpha \prec \beta$, so the attack from α to β will not become a defeat and appears in the abstract argumentation framework; if we make p_2 strictly takes precedence over p_1, then we might get $\beta \prec \alpha$, and the attack from β to α does not become a defeat. Moreover, assuming there is a conflict between protecting the autonomous vehicle and protecting dogs at the same time, the addition of $Inanimate(AV)$ and r_4, combined with the expected priority ordering of legal principles ($p_1 < p_2$ more likely), can make the argument for adopting the action "$Protect(AV)$" be strictly defeated.

3.2.2 Credulous justification

An argument α is said to be credulously accepted/justified under certain argumentation if and only if there exists at least one extension that includes α [1]. We say that an action a is credulously justified if and only if there exists a credulously accepted argument α, such that $\mathtt{Conc}(\alpha) = a$.

If an argument α is credulously accepted, then based on Definition 3.4, α is included in at least one preferred extension $E \in Pr$ (denoted as E_{Pr}), obtained by the same abstract argumentation framework.

Then we have the following proposition.

Proposition 3.8 *Let $LeSAC = (\mathcal{L}, \mathcal{R}^T, n, \mathcal{K}^A)$ be an argumentation theory, $\mathcal{A}$ all the arguments constructed by LeSAC, and α an argument. An action a is credulously justified under complete semantics, iff*

(i) $\exists a_1, a_2, \ldots, a_j, \ldots, a_k, \ldots a_n \in \mathcal{K}^A$ *and rules* $r_1, \ldots, r_n \in \mathcal{R}^T$ *such that* $r_1 = a_1, a_2, \ldots, a_j \dashrightarrow a_1'$, $r_2 = a_1', \ldots, a_k \dashrightarrow a_2'$, ..., *and* $r_n = a_1', a_2', \ldots, a_n \dashrightarrow a$*;*

(ii) $\forall \alpha' \in \text{Sub}(\alpha)$, $\nexists \alpha'' \in \text{Sub}(\alpha)$ such that α'' attacks α';

(iii) $\forall \alpha' \in \text{Sub}(\alpha)$, if $\exists \beta \in \mathcal{A}$ such that β attacks α' and $\beta \nprec \alpha'$, then $\exists \gamma \in E_{Pr}$ such that γ attacks β on $\beta' \in \text{Sub}(\beta)$, and $\beta' \prec \gamma$.

The first and second conditions of Proposition 3.8 are the same as Proposition 3.6, while the difference lies in the third condition: for a certain action to be credulously justified, it is sufficient to make the set of arguments and subarguments from which the action is derived to be the subset of a maximal complete set w.r.t. set-inclusion (i.e., a preferred extension).

Due to space limitations, we omit the similar case-based discussion related to credulous justification.

4 Conclusion

This paper preliminarily delineates the conditions under which a certain action can be sceptically or credulously justified with structured argumentation theory at the foundational level of knowledge bases and rules. Exploration from this direction is crucial for understanding how arguments can be effectively enforced within argumentation frameworks, and structured argumentation can offer a more accessible approach for modelling reasoning processes encountered in real-life scenarios. By integrating these conditions with the definition of attack and defeat relation, containing inherent preferences, we have also shed light on the potential for preference updates in the argumentation theory.

Particularly, the discussion on updating argumentation theory can be applied to the *LeSAC* system designed for advancing the design of autonomous vehicles, which we introduced in our previous papers [16,17,15].

For future work, our next steps include: 1. Further discussing the details of the update of argumentation theory and how to determine which update is the most efficient. Much related work on the dynamics and updating of abstract argumentative frameworks can provide ideas for exploration at the abstract level, such as [3,2,11]. 2. In the current work, we consider the situation under the most basic complete semantics in classical argumentation semantics. We will consider more argumentation semantics in the future, define semantics that are more suitable for legal contexts, and discuss how to enforce conclusions under such argumentation semantics.

Appendix

Proofs for Proposition 3.6 and Proposition 3.8.

Proof. [Proposition 3.6]

 ($\Rightarrow$)

(i) Assume a is justified. This implies the existence of an argument α for which $\text{Conc}(\alpha) = a$. Consequently, there must exist premises $\text{Prem}(\alpha) = a_1, a_2, \ldots, a_n \subseteq \mathcal{K}^A$, connected to the conclusion a via applicable inference rules.

(ii) Consider the case $\exists \alpha', \alpha'' \in \text{Sub}(\alpha)$ such that α'' attacks α'. According to

Definition 3.3, $\exists \beta \in \mathtt{Sub}(\alpha)$ such that α'' attacks α' on it, and in either case of attacks, β counterattacks α''. Then according to the definition of defeats, as long as the preference between arguments is reasonable, this leads to either β defeating α'' or vice versa. Therefore there is a subargument of α that defeats α. Then according to Definition 3.4, if α is defended under a complete extension E, then there is an argument in E that defeats this subargument, thereby defeating α, which contradicts that E is conflict-free.

(iii) Assume the opposite. β defeats α on α'. If such a γ does not exist, then α is not defended by the grounded extension, hence α is not included in all complete extensions, contradicting that $\mathtt{Conc}(\alpha) = a$ is sceptically justified under the complete semantics.

($\Leftarrow$) By (i) an argument α can be constructed such that $\mathtt{Conc}(\alpha) = a$; by (ii) $\mathtt{Sub}(\alpha)$ is conflict-free; and by (iii) all $\alpha' \in \mathtt{Sub}(\alpha)$ are defended by a grounded extension E_{Gr}. According to Definition 3.4, a grounded extension is complete, hence E_{Gr} includes all elements of $\mathtt{Sub}(\alpha)$. According to [1], the grounded extension is included in any complete extension, therefore $\mathtt{Conc}(\alpha) = a$ is sceptically justified under the complete semantics. □

Proof. [Proposition 3.8]

($\Rightarrow$) The proof for conditions (i) and (ii) is the same as in Proposition 3.6. For condition (iii), suppose the opposite. If such a γ does not exist, then since a preferred extension is a maximal complete extensions w.r.t. set-inclusion, α is not defended by any complete extensions, therefore not included in any complete extensions, contradicting that $\mathtt{Conc}(\alpha) = a$ is credulously justified under complete semantics. This contradiction establishes the necessity of γ's existence for the credulous justification of a under the complete semantics.

($\Leftarrow$) By (i) an argument α can be constructed such that $\mathtt{Conc}(\alpha) = a$; by (ii) $\mathtt{Sub}(\alpha)$ is conflict-free; and by (iii) all $\alpha' \in \mathtt{Sub}(\alpha)$ are defended by a preferred extension E_{Pr}. By replacing E_{Gr} with E_{Pr} in the proof of Proposition 3.6, we can prove that $\mathtt{Sub}(\alpha) \cup E_{Pr}$ is conflict-free, then $\mathtt{Sub}(\alpha) \cup E_{Pr}$ is at least a subset of a complete extension. Therefore, $\mathtt{Conc}(\alpha) = a$ is credulously justified under the complete semantics. □

References

[1] Baroni, P., M. Caminada and M. Giacomin, *An introduction to argumentation semantics*, The Knowledge Engineering Review **26** (2011), p. 365–410.

[2] Baumann, R., *What does it take to enforce an argument? minimal change in abstract argumentation*, in: *Proceedings of the 20th European Conference on Artificial Intelligence (ECAI 2012)* (2012), pp. 127–132.

[3] Baumann, R. and G. Brewka, *Expanding argumentation frameworks: Enforcing and monotonicity results*, in: *Proceedings of the 2010 Conference on Computational Models of Argument: Proceedings of COMMA 2010* (2010), p. 75–86.

[4] Borg, A. and F. Bex, *Enforcing Sets of Formulas in Structured Argumentation*, in: *Proceedings of the 18th International Conference on Principles of Knowledge Representation and Reasoning (KR'21)*, 2021, pp. 130–140.

[5] Bundy, A. and X. Li, *Representational change is integral to reasoning*, Philosophical Transactions of the Royal Society A **381** (2023), p. 20220052.

[6] Caminada, M. and L. Amgoud, *On the evaluation of argumentation formalisms*, Artificial Intelligence **171** (2007), pp. 286–310.

[7] Dung, P. M., *On the acceptability of arguments and its fundamental role in nonmonotonic reasoning, logic programming and n-person games*, Artificial Intelligence **77** (1995), pp. 321 – 357.

[8] Li, X., "Automating the repair of faulty logical theories," The University of Edinburgh, 2021.

[9] Li, X. and A. Bundy, *An overview of the abc repair system for datalog-like theories*, in: *The 3rd International Workshop on Human-Like Computing 2022* (2022), pp. 11–17.

[10] Liao, B., M. Anderson and S. L. Anderson, *Representation, justification, and explanation in a value-driven agent: an argumentation-based approach*, AI and Ethics **1** (2021), pp. 5–19.

[11] Liao, B., L. Jin and R. C. Koons, *Dynamics of argumentation systems: A division-based method*, Artificial Intelligence **175** (2011), pp. 1790 – 1814.

[12] Liao, B., M. Slavkovik and L. W. N. van der Torre, *Building jiminy cricket: An architecture for moral agreements among stakeholders*, in: *Proceedings of the 2019 AAAI/ACM Conference on AI, Ethics, and Society, AIES 2019*, Honolulu, HI, USA, 2019, pp. 147–153.

[13] Lu, Y., Y. Lin, X. Li, A. Bundy, B. Schafer and A. Ireland, *Logic and theory repair in legal modification*, in: *Proceedings of the 3rd International Joint Conference on Learning & Reasoning (IJCLR): CogAI 2023*, 2023.

[14] Lu, Y. and Z. Yu, *Argumentation theory for reasoning with inconsistent ontologies*, in: *Proceedings of the 33rd International Workshop on Description Logics (DL 2020)*, 2020.

[15] Lu, Y., Z. Yu, Y. Lin, B. Schafer, A. Ireland and L. Urquhart, *Handling inconsistent and uncertain legal reasoning for ai vehicles design*, in: *Proceedings of Workshop on Methodologies for Translating Legal Norms into Formal Representations (LN2FR 2022)*, 2022, pp. 76–89.

[16] Lu, Y., Z. Yu, Y. Lin, B. Schafer, A. Ireland and L. Urquhart, *A legal support system based on legal interpretation schemes for ai vehicle designing*, in: *Proceedings of the 35th International Conference on Legal Knowledge and Information Systems (JURIX 2022)* (2022), pp. 213–218.

[17] Lu, Y., Z. Yu, Y. Lin, B. Schafer, A. Ireland and L. Urquhart, *A legal system to modify autonomous vehicle designs in transnational contexts*, in: *Proceedings of the 36th International Conference on Legal Knowledge and Information Systems (JURIX 2023)* (2023), pp. 347–352.

[18] Modgil, S. and H. Prakken, *A general account of argumentation with preferences*, Artificial Intelligence **195** (2013), pp. 361–397.

[19] Prakken, H., *An abstract framework for argumentation with structured arguments*, Argument & Computation **1** (2010), pp. 93–124.

[20] Rapberger, A. and M. Ulbricht, *On Dynamics in Structured Argumentation Formalisms*, in: *Proceedings of the 19th International Conference on Principles of Knowledge Representation and Reasoning*, 2022, pp. 288–298.

[21] Rosenfeld, M., *Dworkin and the one law principle: A pluralist critique*, Revue Internationale de Philosophie **59** (2005), pp. 363–392.

[22] Xu, K., B. Liao and H. Huang, *Updating argumentation frameworks for enforcing extensions*, in: B. Liao, T. Ågotnes and Y. N. Wang, editors, *Dynamics, Uncertainty and Reasoning (CLAR 2018)* (2019), pp. 63–79.

Towards a Logical Analysis of Epistemic Injustice

Joris Hulstijn Huimin Dong Réka Markovich

University of Luxembourg, Esch-sur-Alzette, Luxembourg
joris.hulstijn@uni.lu, huimin.dong@uni.lu, reka.markovich@uni.lu

Abstract

Cases of bias and unfair decisions in automated decision-making are heavily discussed. When unfair decision outcomes can be attributed to an unjust difference in the knowledge of various groups of subjects, we can speak of epistemic injustice. In this paper, we analyze various kinds of epistemic injustice, such as testimonial, hermeneutical, distributional, and content-focused epistemic injustice, and show how they can be conceptualized. We then provide a formalization of the difference in group knowledge, in a version of epistemic logic. After that, we discuss two cases of badly designed information systems for government decision-making. We analyze key observations from the cases pointing out that they constitute a form of epistemic injustice.

Keywords: social epistemology, injustice, bias, epistemic logic

1 Introduction

Automated decision-making systems (will) make decisions that matter. An important concern about applications of AI is bias in decision-making [18, 26]. Systems, as well as humans who are supported by (or provide the data to such) systems, do make decisions that are unfair or unjustified for certain groups, relative to other groups. Some groups in society do not know the procedures or are unable to fill out the required forms [24]. Therefore, these groups are treated unfairly. For example, some groups are relatively more likely to get into trouble with the tax office than other groups [10, 19]. How can we analyze such cases? In general, errors in automated decision-making may occur because (i) the algorithm or decision rules are biased, (ii) the data set on which the system was trained is biased, or (iii) the human operator who should counterbalance possible system bias, is not supported to do this difficult task [18].

Social epistemology investigates the epistemic aspects of social interactions [17]. Fricker has proposed to analyze some of the above-mentioned cases in terms of *epistemic injustice* [15]. The term has two parts: (i) *injustice*: there is a moral wrong or a legal right that is violated (ii) *epistemic*: the wrong is based on a difference in the knowledge or information that is available to some groups in society relative to other groups [15, 12, 13, 20, 22]. In this paper, we provide an initial formal conceptual analysis of the epistemic part, showing that in some cases (types) the injustice is due to the decision-maker's epistemic state (regarding the subject group's knowledge).

The discussion about epistemic injustice is part of a wider trend combining ethics and epistemology, for instance, in business ethics [12] and medical ethics [6]. Here we look at another application: government decision-making, e.g. [24]. We refer to observations from two scandals about errors in government decision-making: the 'Toeslagenaffaire' [10] in the Netherlands, and the RoboDebt case in Australia [19].

The aim of the paper is to: (i) identify the main types of epistemic injustice from the literature, and provide a possible explanation for the mechanisms that give rise to epistemic injustice; (ii) use epistemic logic to specify the difference in knowledge between groups; (iii) explain observations from the cases [10, 19], to establish whether those constitute a form of epistemic injustice.

The paper is structured as follows. In Section 2 we analyze the notion of epistemic injustice. In Section 3 we formalize the definition in an epistemic logic and in Section 4 we provide an initial formal analysis of some types of epistemic injustice. Section 5 describes the cases, and provides observations that illustrate epistemic injustice. The paper ends with conclusions and suggestions for further research.

2   What is Epistemic Injustice?

Epistemic injustice is a form of injustice related to knowledge. The Handbook of Epistemic Injustice says "epistemic injustice refers to those forms of unfair treatment that relate to issues of knowledge, understanding, and participation in communicative practices." [22, preface]. The notions of *justice* and *injustice* have been widely discussed in moral and legal philosophy. We do not recall this literature here, but focus only on the epistemic part of epistemic injustice, based on how it is handled in social epistemology. In the context of social epistemology [22], we look at differences between groups in society. So, injustices are studied that can be attributed to a (lack of) knowledge in one group, relative to another. As we will point out, in several cases, not the actual knowledge of one agent or group is what is relevant, but the beliefs—or prejudices—of others regarding it.

2.1   Types of Epistemic Injustice

The notion of epistemic injustice was orginally introduced by Fricker [15]. In this paper, we distinguish four types: (C1) distributional injustice, (C2) testimonial injustice, (C3) hermeneutic injustice, and (C4) content-focused injustice. Fricker [15] discusses testimonial and hermeneutic epistemic injustice the most, but distributional injustice is also mentioned, and worked out in more detail in a later work of her [16]. Content-focused epistemic injustice was introduced much later by Dembroff and Whitecomb [14], although such cases were known before.

Distributional injustice concerns a situation in which one group has less (access to) knowledge than another (privileged) group, leading to unfair treatment. For lack of a better word, we will call this *distributional injustice*, because it is based on an unequal distribution of knowledge or education [16, p 1317]. We should emphasize however, that not all cases of unequal distribution of knowledge constitute an injustice. A large part of society is based on merit and education. People who study and know more get more opportunities. It only becomes an injustice when access to knowledge is withheld from certain groups. This may be, for example, when documents are not made available in a minority language.

Testimonial injustice is a form of unfairness related to the relative trust in some-

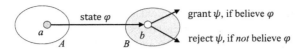

Fig. 1. Basic setting: speaker a from group A makes assertion φ to addressee b from group B. Speaker a *depends* on a decision by b to receive some benefit ψ. Decision to grant ψ requires that b believe φ. Members of group B are *prejudiced* against A.

one's capacity as a knower, for example, as an expert witness in a trial. An injustice of this kind can occur when someone is not believed or even ignored, because of group properties like their sex, sexuality, gender presentation, race, disability, or more generally, because of their identity [15]. For example, when Marilyn Von Savant, the person having the highest recorded IQ in the world, provided the *correct* answer to the Monty Hall problem in the column of Parade Magazine, tens of thousands of people (including many mathematicians and other academics) reacted by publicly rejecting *harshly* what Von Savant said. Most of the reactions just considered her answer *unimaginable* to be correct, which—as we will see—falls into the category of content-focused injustice (C4), but several reactions reflected upon her being a woman as a reason for being wrong, which is a case of testimonial injustice (C2). [1]

Hermeneutical injustice is a form of unfairness related to how people interpret and understand their situation, and their ability to formulate believable statements about that situation. The term 'hermeneutic' comes from the Greek word for interpreter. After all, hermeneutics is the philosophical discipline that is concerned with interpretation or understanding. An example given by Fricker is that before the 1970s, victims of sexual harassment had trouble describing in court the behavior of which they were the victims, because the concept had not yet been articulated. In particular, legal procedures demanded physical evidence of abuse, which was hard to obtain.

Content-focused epistemic injustice [14] is a form where the unfair treatment is not based on characteristics of the group but rather on the content of what they say. A particular type of message (content) is ignored or mistaken, for example, because it doesn't align with the consensus opinion in the dominant group.

2.2 Setting

Consider the following setting, characterized by *prejudice* and *dependence* (Figure 1) There are two groups: A and B, whose members speak the same language, but have different knowledge and a different ontology to conceptualize the world.

Suppose speaker $a \in A$ makes assertion φ to addressee $b \in B$, because as part of his tasks b must make a decision ψ on which a depends. For example, b must decide to grant a subsidy or not. Formally, the decision depends on b accepting a's statement φ as true. Statement φ is typically supported by documents of written evidence (financial statements; tax returns etc). For example, when a doesn't provide proof of residence in the municipality in which the subsidy is claimed, b will not believe that a is eligible for a subsidy, and b will not grant the subsidy.

---

[1] See: https://priceonomics.com/the-time-everyone-corrected-the-worlds-smartest/

2.3 Speech act theory

We will now analyze the social setting presented in Figure 1, in terms of Speech Act theory [2, 25]. In Table 1 we analyze the utterance in several layers: form (syntax), content (semantics), and function (pragmatics). At each layer a performs various acts: (1a) locutionary: sending a statement message, (2a) illocutionary: conceptualizing and encoding the statement, and (3a) perlocutionary: making a statement as part of a valid decision request. These acts from a are complemented by corresponding acts by b, at each level: (1b) receiving the statement, (2b) understanding the meaning of the statement, and (3b) accepting the statement as valid. This reflects the idea that communication is a joint action by speaker and addressee, at various layers [9] .

If we say "b doesn't believe a's statement φ", this may have several possible reasons. There are in fact six potential points in the model, where something may go wrong. Now the four types of epistemic injustice can be positioned in this model.

C1 **Distributional**. Members of A lack knowledge φ, which members of B do have. Knowledge of φ is necessary to obtain some benefit ψ. So a cannot make a statement that φ (1a) that is received by b (1b), by lack of preparatory conditions.

C2 **Testimonial**. Members of A are in general not seen as trustworthy, about topics related to φ. Therefore, b doesn't accept statement φ as true (3b), by a supposed lack of sincerity of a.

C3 **Hermeneutical**. Members of A lack knowledge, to conceptualize and encode the intended meaning φ in a statement, that b will accept. (2a)

C4 **Content-focused**. Members of B share a consensus that φ is false. Therefore, content φ will not be understood by b (2b), or a's statement φ will not be accepted by b to be true (3b).

For content-focused epistemic injustice, there are two possible explanations. (i) Confirmation bias: φ may not be accepted as true, to protect the group consensus (3b). If an individual b accepted φ, he would threaten the group's consensus or place himself outside of the group. (ii) Cognitive dissonance: statement φ is so far removed from what is considered normal, that for b it takes much more effort to process and accept it, than to reject it. For example, b lacks the ontology to understand the problems of a (2b). This seems to be the counterpart of hermeneutical injustice for the addressee. Sometimes, these types of epistemic injustice strengthen each other. For example, suppose b believes that φ must be false because it goes against the consensus (content-focused). So for b, the reason that a makes a false statement φ can only be that a seeks to gain from it. So b doubts the sincerity of a (testimonial injustice).

	a		b
1. locutionary:	sending a statement	–	receiving the statement
2. illocutionary:	conceptualizing and encoding the meaning	–	understanding and decoding the meaning
3. perlocutionary:	making a valid statement	–	accepting the statement as valid

Table 1

Model of information exchange as joint action at various linguistic levels [2, 9]

2.4 Example: Complaints

The two cases we will discuss in Section 5 below [10, 19], are both prescriptive systems: they have a word-to-world direction of fit. In case of a conflict about the contents of the system, the subjects of the decision are at a disadvantage. To file a complaint, they have to substantiate the claim by evidence, but, by the nature of the system, often there is no credible source to testify their version of reality. That means that a conflict becomes a game of trust. Now consider a subject to an automated decision, who feels she is being wronged. That means she must file a complaint. However, in both cases [10, 19], complaints were ignored or rejected without motivation.

The four types of epistemic injustice we identified above based on [15, 14] can be used to explain the social mechanisms why complaints were ignored. In some cases, more categories apply. For example, a conflict of opinion (C4) may further reduce the perceived sincerity of the group (C2).

C1 *Distributional injustice*: subjects who want to complain about a system error, lack technical knowledge about the system, lack legal knowledge of the grounds for the decision, or lack knowledge of the procedures to file a complaint, and do not know people who could help them.

C2 *Testimonial injustice*: subjects who complain are not believed by the officials, because they are part of a specific group. Or, the officials who treat the complaint demand documented evidence, of a kind that is not available for this group (e.g. proof of income). Usually, subjects are expert on their own situation. So here they are wronged in their capacity of knowing the relevant facts that matter to the case.

C3 *Hermeneutical injustice*: subjects who want to complain about a system error, know very well that something is wrong. But they are not experts in tax law and do not know the formal legal terminology (e.g. tax debt; evidence) in terms of which to analyze their own situation and formulate a specific complaint.

C4 *Content-focused injustice*: subjects who file a complaint about the system, thereby claim that the system is wrong. The people responsible for the system, believe the system cannot be wrong. After all, "the system has been carefully developed, and has been tested by experts, etc" The complaint is much harder to understand, than the alternative (cognitive dissonance). Moreover, the complaint contradicts the consensus opinion among the system experts. People will not actively seek evidence to disproof that consensus (confirmation bias). That means, that the complaint must be wrong or even insincere.

3 Epistemic Injustice: An Epistemic Logic Perspective

The different types of epistemic injustice can be analyzed in a general model of knowledge and information exchange which is the basis for our initial logical analysis. We will now discuss a version of epistemic logic to provide such a model. But epistemic injustice goes beyond the wrongful recognition of an individual's epistemic status, it also examines how this misrecognition can lead to unfairness, which leads us to rely on logical considerations of actions as well since, in this paper, we conceptualize this unfairness as unsatisfactory decisions for a person making a request for *action*. In this section, thus, we will explore both the epistemic and action-oriented elements of epistemic injustice by introducing an action-based epistemic logic.

In this logic, the language we use to address the types of epistemic injustice includes *individual* knowledge K_a, *individual* belief B_a, and *common* belief C_G. We also include the modality E_a to illustrate *actions* or decisions.

Definition 1 (Language) Let *Prop* be a countable set of atomic propositions, and $\mathscr{I}$ be a finite set of agents. The language $\mathscr{L}$ is defined as follows:

$$\varphi ::= p \mid \neg\varphi \mid \varphi \wedge \varphi \mid K_a\varphi \mid B_a\varphi \mid C_G\varphi \mid E_a\varphi \mid \Box\varphi,$$

where $p \in Prop$, $a \in \mathscr{I}$, and $\emptyset \neq G \subseteq \mathscr{I}$.

The dual of K_a, denoted as $\hat{K}_a$, illustrates the consistency with agent a's knowledge. So $\hat{K}_a\varphi$ is defined as $\neg K_a\neg\varphi$ and read as "φ is consistent with agent a's knowledge." Similarly, the dual of belief $\hat{B}_a\varphi$ is defined as $\neg B_a\neg\varphi$ and read as "φ is consistent with agent a's belief," and the dual of universal modality $\Box$ is the existential modality $\Diamond$ such that $\Diamond\varphi := \neg\Box\neg\varphi$. The B-modality is a KD4-modality, K- and $\Box$-modalities are S5-modalities, and C_G-modality is a KD4-modality.

The sender's or applicant's request is dealt with by the decision maker, she has the right to decide whether to fulfill the request or not. To express the fulfillment of the right of decision, our language includes an additional operator, denoted as E_b, to represent agent b's action execution. Thus, $E_b\psi$ can be interpreted as "Agent b takes specific actions to ensure that ψ holds true," or, from the perspective of agency theory or STIT logic [7, 4], "Agent b ensures that ψ is the case." In this context, we adopt Chellas's proposal [8] and treat this action operator, E_b, simply as a T-operator. The dual of E_b is denoted as $\hat{E}_b$, and $\hat{E}_b\psi$ is equal to $\neg E_b\neg\psi$.

Definition 2 (Models) A structure $M = (W, \{R_a\}_{a\in\mathscr{I}}, \{D_a\}_{a\in\mathscr{I}}, \{\sim_a\}_{a\in\mathscr{I}}, V)$ is a model when it satisfies the following conditions:

- W is a non-empty set of states;
- R_a is an equivalence relation over W;
- D_a is a transitive and serial relation over W, such that $D_a \subseteq R_a$;
- $\sim_a$ is an equivalence relation over W;
- $V : Prop \to \wp(W)$ is a valuation function.

The accessibility relation R_a interprets individual a's knowledge, D_a represents individual a's belief, and $\sim_a$ represents individual a's ability to execute actions. We can define the transitive closure of all individuals' beliefs in the group G as $D_G = (\bigcup_{a\in\mathscr{I}} D_a)^+$. Now our modalities can be interpreted as usual:

$$M, w \models K_a\varphi \text{ iff } R_a[w] \subseteq ||\varphi||, \qquad M, w \models E_a\varphi \text{ iff } \sim_a [w] \subseteq ||\varphi||,$$
$$M, w \models B_a\varphi \text{ iff } D_a[w] \subseteq ||\varphi||, \qquad M, w \models \Box\varphi \text{ iff } W \subseteq ||\varphi||,$$
$$M, w \models C_G\varphi \text{ iff } D_G[w] \subseteq ||\varphi||,$$

where $||\varphi|| = \{w \in W \mid M, w \models \varphi\}$. So the following statements are valid:

CB $\quad C_G\varphi \to B_a\varphi$ if $a \in G$; $\qquad\qquad$ KB $\quad K_a\varphi \to B_a\varphi$.

By applying this action-based epistemic logic, we are able to define, for instance, the epistemic aspects of Fricker's notion of testimonial injustice, which involve the incorrect recognition of one's capacity as a knower.

- Agent b wrongly recognizes agent a's knowledge: $K_a\varphi \wedge B_b\neg K_a\varphi$;
- Agent b wrongly recognizes agent a's credibility of knowledge: $K_a\varphi \wedge \neg B_b K_a\varphi$.

In the next section, we will explore the formalization of the four types of epistemic injustice within our epistemic logic.

4 Towards a Formal Theory of Epistemic Injustice

We present four assumptions, (Ai) – (Aiv), that characterize the situation of a decision maker b, who, as recipient of a statement φ, doubts the credibility of the sender a and subsequently rejects b's request, especially when sender $a \in A$ is perceived to be outside the privileged group B, and $b \in B$ (see Figure 1). We introduce notation $A \leqslant B$ to indicate that group A holds a disadvantaged epistemic position relative to group B. In this paper, we have not introduced a semantics for the expressions such as $A \leqslant B$. However, it's worth noting that this is a viable possibility [2] [3]. The framework conditions provided here offer valuable insights for undertaking this task.

Within this context, $a \in A$ represents the message sender, while $b \in B$ signifies the message receiver and decision maker. In addition, φ represents the evidence submitted by a and ψ represents the requested decision to be made by b. Proposition $\Box(\psi \to \varphi)$ represents that "Evidence φ submitted by a is required to fulfill the request for ψ". This means that the evidence is a necessary condition for fulfilling the request. The formalization of the four assumptions are as follows: [4]

Ai For all $b \in B$: $\Box(\psi \to \varphi) \to (E_b\psi \to B_b\varphi)$;

Aii For all $a \in A$ and $b \in B$ with $A \leqslant B$: $B_b\varphi \to K_a\varphi$;

Aiii For all $a \in A$ and $b \in B$ with $A \leqslant B$: $B_b\varphi \to B_b K_a\varphi$;

Aiv For all $b \in B$: $C_B\neg\varphi \to B_b\neg\hat{E}_b C_B\varphi$.

These assumptions serve to elucidate the underlying factors leading to *prejudice* against individuals in disadvantaged positions by those in privileged positions. Prejudice, as delineated by these four assumptions, is not solely a manifestation of power

---

[2] Intuitively, $A \leqslant B$ when according to any $b \in B$, any $a \in A$ believes fewer formulas than $b \in B$. Let $\mathsf{Form}_w(b,a) = \{\varphi \mid M,w \models B_b K_a\varphi\}$. We have $M,w \models A \leqslant B$ iff $\mathsf{Form}_w(b,a) \subseteq \mathsf{Form}_w(b,b)$ for all $a \in A, b \in B$.

[3] In this setting, notation $A \leqslant B$ is relative to the topic area of proposition φ. Suppose $Prop$ is divided in overlapping subsets $T \subset Prop$, that denote a topic area, such as finance, or sports, etc. A formula can be classified by the topic of the proposition letters in it. Now in general, a person b trusts a person a to know φ whenever the topic of φ is in the competence areas of person a, according to be b. See [11]

[4] The frame conditions to validate (A1) – (A4) are as follows:

(Ai) $\forall wu \in W(wD_b u \to w \sim_b u)$;

(Aii) $\forall wu \in W(wD_b u \to wR_a u)$, if $A \leqslant B$, $a \in A$ and $b \in B$;

(Aiii) $\forall wuv \in W(wD_b u \wedge uR_a v \to wD_b v)$, if $A \leqslant B$, $a \in A$ and $b \in B$;

(Aiv) $\forall wuv \in W(wD_b u \wedge u \sim_b v \to wD_b v)$.

owned by privileged decision-makers. It also arises from the presence of several *irrational* assumptions underlying their decision-making processes. These irrationalities become apparent in the assumptions we have outlined above:

Assumption (Ai) captures the decision rights of b to grant ψ. Here $E_b \psi \rightarrow B_b \varphi$ means that b believes evidence φ is a necessary condition for b to ensure ψ.

Assumption (Aii) reflects the irrationality of the advantageous and dominant position held by the receiver: If the decision maker $b \in B$, positioned in an advantaged state B (which is illustrated as $A \leqslant B$ [5]), and holds a certain belief, it serves as a compelling rationale to posit that members of the disadvantaged group must possess the same knowledge. In essence, the beliefs of the privileged party take precedence over the knowledge of the disadvantaged party. Further, the concept of *prejudice* is exemplified by this interdependence: When an individual lacks knowledge of φ, this becomes a reason that the information sent by this agent is not believed by the decision maker, primarily due to their skepticism toward the disadvantaged group (i.e. $A \leqslant B$).

Assumption (Aiii) highlights the dominant position of the advantaged group from a different perspective. When an individual within the advantaged group accepts a piece of information, they believe that any member in the disadvantaged group must possess this information as their knowledge. This phenomenon underscores the concept of "*prejudice*" as one type of interdependence of communication: The beliefs of the dominant individual influence their perceptions regarding the knowledge of those in the disadvantaged group.

Assumption (Aiv) sheds light on the concept of common ground within the privileged group. Simply speaking, when a piece of information is established as part of the common ground for the group, every individual within that group believes it is impossible to revise such a common belief.

To understand epistemic injustice, these four assumptions play a pivotal role, as outlined in Table 2. Note that the reasoning behind distributional injustice, testimonial injustice, and content-focused injustice differs in three key aspects, respectively: distinctions in factual information, beliefs of decision-makers about credibility of groups, and beliefs of decision-makers about credibility of content.

Distributional injustice is rooted in the fact that the sender a, who is in a disadvantaged position, lacks knowledge of the evidence φ, which can be expressed as $\neg K_a \varphi$. Given this fact and the communication assumption (Aii), it leads to "*weak belief*": the decision maker b does not believe in the evidence φ submitted by agent a, denoted as $\neg B_b \varphi$. This type of belief is considered *weak*, because it is derived from a basis of the other's lack of knowledge [27, 21]. Following assumption (Ai), which addresses the decision right of b, the decision doesn't fulfill the request ψ from agent a: $\neg E_b \psi$.

In contrast, the reasoning process for testimonial injustice follows a different path. While it also involves a weak belief $\neg B_b \varphi$, it is inferred from a distinct basis of information. Testimonial injustice is rooted in the fact that the sender a indeed possesses

---

[5] In this paper, we introduce a binary relation denoted as $\leqslant$ to illustrate intergroup *prejudice*, providing a simplified representation for this key concept in epistemic injustice. While it's acknowledged that prejudice in the real world can be influenced additionally by various factors, such as topics that are discussed in [11], our current focus centers on establishing the logical principles for defining prejudice between groups. The examination of prejudice with respect to both groups and topics remains a subject for future research.

knowledge of the evidence φ (i.e., $K_a\varphi$). It also relies on the epistemology assumption that the decision-maker does believe the sender genuinely lacks knowledge of the evidence, denoted as $B_b\neg K_a\varphi$. This belief is referred to as "*strong belief*," because it is assumed and not derived. This strong belief is labeled as a *prejudice*, because it presupposes that **everyone** in a disadvantaged group A lacks knowledge about this topic area, regardless of its actual veracity. From this strong belief, assumption A2 and axiom D, the weak belief $\neg B_b\varphi$ can also be inferred. Ultimately, assumption A3 leads to non-fulfillment of the request.

	Facts	Beliefs	Request Fulfillment	Inferential Elements
Distributive Injustice	$\Box(\psi \to \varphi)$ $\neg K_a\Box(\psi \to \varphi)$ $\neg K_a\varphi$	Weak: $\neg B_b\varphi$	$\neg E_b\psi$	Aii, $\neg K_a\varphi$ Ai, $\Box(\psi \to \varphi)$, Weak
Testimonial Injustice	$\Box(\psi \to \varphi)$ $K_a\Box(\psi \to \varphi)$ $K_a\varphi$	Strong: $B_b\neg K_a\varphi$ Weak: $\neg B_b\varphi$	$\neg E_b\psi$	Aiii, D, Strong Ai, Weak
Content-focused Injustice	$\Box(\psi \to \varphi)$ $K_a\Box(\psi \to \varphi)$ $K_a\varphi$	Strong: $C_B\neg\varphi$ Weak$_1$: $C_B\neg K_a\varphi$ Weak$_2$: $B_b\neg K_a\varphi$ Weak$_3$: $\neg B_b\varphi$ Weak$_4$: $B_b\neg\hat{E}_bC_B\varphi$	$\neg E_b\psi$	T, NEC$_C$, Strong CB, Weak$_1$ Aiii, D, Weak$_2$ Ai, Weak$_3$ Aiv, Strong

Table 2
A Classification of epistemic injustice (C1,C2,C4), where $a \in A$ and $b \in B$ with $A \leqslant B$.

In the last row of Table 2, content-focused injustice, we can model two cases: (i) the individual case $B_b\neg\varphi$ and therefore $\neg B_b\varphi$, so the request is rejected, and also $B_b\neg K_a\varphi$, so the requester is denied in her right as a knower, and (ii) the group consensus case, $C_B\neg\varphi$ and therefore $\neg C_B\varphi$, so the request would be rejected by any official, but also $C_B\neg K_a\varphi$, so the requester is by consensus denied in the right as a knower.

Note however, that in this simple form of epistemic logic we cannot distinguish between the failure of b to understand φ, and the failure of b to publicly accept statement φ. We cannot express the need for belief revision either, if φ would be accepted as true in case $\neg\varphi$ is already believed, which would take effort. For a similar reason, the current formalism lacks the tools to address (C3) hermeneutical injustice, as we cannot express that only some agents have an ontology to understand (for b) or to express (for a) a problem situation, so we leave all these to future work. What we can represent is that b would be inclined to believe that φ based on a statement by a, but still believes that it would be impossible to convince the others: $C_B\neg\varphi$ but $B_b\Diamond\varphi$ (while, let's say, most of the group members also believe that it is actually $\Box\neg\varphi$), so also $B_b\Diamond K_a\varphi$, but still $B_b\neg\Diamond C_B\varphi$ (or just $B_b\neg\Diamond E_bC_B\varphi$) so he doesn't do anything.

5 Cases

In this section we analyze two cases of automated decision making for government, that display errors in decision making. The purpose is to test if these errors can be classified as by the four types of epistemic injustice. We use publicly available sources, in particular parliamentary investigation reports [10, 1, 19].

5.1 Case 1. Toeslagenaffaire (Netherlands)

The Toeslagenaffaire (child care benefits scandal) is a complex and sensitive set of interrelated cases and problems, of a political, legal, technical and administrative nature, in the Netherlands in the period 2010-2017 [10]. The consequences are still felt. Here we can only provide a few telling observations. Observations are **bold** in the text.

The benefit scheme started with politicians' wish stimulate women to get paid work and create a market for childcare. The state is funding care centers indirectly by reimbursing parents for the costs incurred. This may involve several hundreds of Euros per month. To get reimbursed, parents have to apply for childcare benefits. Childcare benefit is a conditional entitlement. Parents are only entitled to a certain amount of benefit, if they actually use childcare for a certain number of hours, if the childcare center is approved, and if their combined income stays below a certain threshold.

To provide evidence of these conditions, parents have to fill out forms and supply documents, often obtained from other parties. Given the complexity of the forms and rules, it is likely that mistakes are made. Moreover, many people do not know in advance exactly how much income they will earn. In general, social benefit agencies are used to working with such estimates. However, for Toeslagen, the law made the families responsible for providing exact numbers about their situation. This attitude was driven by a political pressure to combat fraud, in the years before the scheme.

The Netherlands Tax Administration was tasked with executing the scheme, in particular the department Benefits (Belastingdienst/Toeslagen). At the time, the tax office was seen as more competent in IT, than other government agencies.

Here we focus on a specific case (CAF-11). The tax office generally applies risk-based supervision [5]. However, here **subjective risk indicators** were used. For example, a person owns an expensive car without the income to support it. Indicators were used to identify suspected people, and place them in a system called CAF-11, which was effectively a blacklist. Indicators were not verified to be useful for finding fraud. Some risk indicators, like double nationality, were later ruled to be ineffective, unnecessary for the task and therefore unlawful (GDPR), and even discriminatory [3].

Originally, the list was only used by fraud teams, to collect early warning signals. From 2014, the list was also used for regular application processing. Being on the list itself became a sufficient reason to be denied childcare benefits. Citizens received **no explanation** for such rejections. Naturally, these citizens complained, but many **complaints were ignored** or **rejected**. This shows a pattern [1]:

"In the CAF 11 case, the focus on fighting fraud caused **institutional bias**, according to the committee. That bias meant that from the outset, the actions of Benefits were based on the suspicion that the CAF 11 parents had committed fraud. A suspicion that was not based on the personal actions of these parents, but on the mere fact that they were being monitored as part of the CAF 11 file." Translation of [1].

The term *institutional bias* (institutionele vooringenomenheid) in the committee report, led to a heated political debate. It means that Benefits was prejudiced against

Observation	Effect	Type
subjective indicators	decision based on group identity, not evidence	testimonial
no explanation	citizens had no knowledge of what went wrong	distributional
complaints ignored	officials had no knowledge of what went wrong	dist. test. herm. cont.
institutional bias	decisions based on systematic prejudice	test.; cont.
aimed at shortcomings	mistakes treated as intentional or grave neglect	testimonial
no opportunity to correct	treated as criminal	herm.; cont.
all-or-nothing	disproportional; treated as criminal	testimonial
no redress mechanism	increased duration, harm	herm.; cont.

Table 3

Observations in the Toeslagenaffaire analyzed as types of epistemic injustice

a group of parents, namely those listed on the CAF 11 file. The bias was institutional because it was given a place in work instructions and also worked through in objection and appeal procedures, in the recovery measures as well as in new applications for childcare benefits from these parents. For instance, hearings revealed that management overseeing the CAF fraud team accepted that if about 80% of the people on the list were indeed 'bad guys', 20% of people on the list must be 'good guys', and thus unjustifiably targeted (80/20 rule). ([10, p 47].

After terminating a benefit, the tax office would investigate whether parents had been entitled to the benefits received. This investigation was deliberately **aimed at finding shortcomings**—even the slightest ones—in administration, payments, or supporting documents to have the benefit withdrawn. Citizens normally get an opportunity to correct their statements, except when the tax office believes manipulations are intentional. However, here citizens were given **no opportunity to correct**. That is, citizens making mistakes were treated as criminals. If an administrative discrepancy was discovered, the entire benefits amount had to be paid back, not just a correction (**all or nothing approach**). This increased the impact on families, with many perceiving it as a punishment. Later, legal scholars agreed this was disproportional [10].

The law that regulates the social benefits scheme, does not give officials explicit discretionary power to deviate from the policies. For example, there is no hardship clause, as is customary in other social benefit schemes. The strict application of the law was also confirmed in case law. The Council of State (Raad van State) often ruled in favour of Toeslagen, in appeal cases in which the strict application of the right to childcare benefits was questioned. Internal doubts by officials in the tax office about the hardship, were silenced, with reference to the ruling by the Council of State. So initially, there were **no mechanisms for redress or appeal**. This greatly extended the duration and impact of the hardships sustained by families.

Table 3 repeats the main observations and shows that they constitute a case of epistemic injustice. To summarize, all forms of epistemic injustice were present in this case. Distributional and testimonial are most common, but also cases of content-focused and hermeneutical were found, especially in the aftermath and the inability of the administration to handle complaints and redress the problems.

5.2 Case 2. Robodebt (Australia)

We can only shortly address the Centrelink case in Australia (New South Wales), known as 'RoboDebt' [19]. As before, observations are shown in bold, and cate-

Observation	Effect	Type
estimates unverified	decision based on estimates, not evidence	testimonial
no explanation	citizens had no knowledge of what went wrong	distributional
complaints ignored	officials had no knowledge of what went wrong	dist. test. herm. cont.
no redress mechanism	increased duration, harm	herm.; cont.

Table 4

Selection of observations in the Robodebt case analyzed as types of epistemic injustice

gorized as forms of epistemic injustice in Table 4. The Centrelink system refers to a long-term project, supported by a computer system, that was meant to automatically calculate and subsequently collect the 'debts' of a citizen. By debts they meant claimed social benefits to which the receiver was not lawfully entitled. The scheme was based on the belief, that fraud with social benefits was widespread. The debt collection was to generate revenues for the government, and earn back the investments.

These reclaimable debts were calculated, based on data about income and social status originally provided by the citizen, and on data transferred from the Australian Tax Office. However, if relevant data was absent, the system would use heuristics and machine learning techniques to **estimate** the missing data, based on the average income of the previous period. We could say this tendency reflects a prejudice about a proper income. Especially for people with fluctuating incomes, this was disadvantageous. Although **unverified**, these estimates were used as evidence to calculate the debt to be paid. Some people had to pay a large amount of debts they didn't owe. Citizens received **no explanation**. Victims found it impossible to understand what they had done wrong, and what caused these large debts. Some victims stated they thought they had lost their minds: it was their word against that of the tax office. **Complaints** were often **ignored or rejected**. Once reports about errors surfaced, it took a long time for the government to admit this. Hence, there was **no redress procedure**.

5.3 Comparison

There are many similarities between the two cases. Both were concerned with social benefits, acclaimed fraud, and a very harsh and strict interpretation of what constitutes financial evidence. Both projects were initiated under large political pressure to get results, to reduce fraud. Both projects showed a large trust in technology from the government, and a lack of understanding for the specific situation of citizens, especially for those in the target audience (people with lower incomes). In both cases, it took a long time for victims to be heard and to get compensation, and eventually, both cases triggered a huge political scandal, and parliamentary committees investigated the legal and political aspects [10, 23, 19].

There are also interesting differences. The Dutch case focused on fraud detection in child care benefits, and was executed by a department of the tax office. The Australian case is about the reclaim of debt (unentitled benefits), based on social security laws. It was executed by a special administrative body, which lacked oversight.

6 Conclusions and future work

We have investigated a complex and sensitive topic from social epistemology: epistemic injustice. We have defined it and identified four types, based on the literature

[15, 16, 14]: distributional, testimonial, hermeneutical, and content-focused epistemic injustice. We have analyzed the mechanisms behind these types using speech act theory, and provided an initial formal characterization of three of these types in an action-based epistemic logic for beliefs and knowledge (distributional, testimonial, and content-focused injustice). We presented four assumptions as foundational elements in the development of a formal theory encompassing these three types. These assumptions elucidate the inferential components at play in the reasoning processes of privileged individuals. There is a lot for future work in the logical representation: e.g., so far, we cannot express the hermeneutical type, we cannot properly express the notion of social power, nor the notion of identity prejudice, which plays a crucial role in Fricker's original work. Currently, this part of the analysis is based on a simplified setting (Figure 1). In the future, we would like to generalize and add a notation for background facts about roles and social relations between agents, that influence epistemic trust and prejudice. We have also analyzed one case about a government decision-making system [10], and list observations corresponding to each epistemic injustice types. In future work, we want to analyze the second case [19] in more detail.

Acknowledgement

This work is supported by the *Fonds National de la Recherche* of Luxembourg through the project Deontic Logic for Epistemic Rights (OPEN O20/14776480).

References

[1] Adviescommissie Uitvoering Toeslagen, *Omzien in verwondering*, Tweede Kamer der Staten Generaal (2019).

[2] Austin, J. L., "How to do things with words," Harvard UP, Cambridge MA, 1962.

[3] Autoriteit Persoonsgegevens, *Belastingdienst/Toeslagen: De verwerking van de nationaliteit van aanvragers van kinderopvangtoeslag*, Report z2018-22445 (2020).

[4] Belnap, N., M. Perloff and M. Xu, "Facing the future: agents and choices in our indeterminist world," Oxford University Press, 2001.

[5] Black, J., *The emergence of risk-based regulation and the new public risk management in the united kingdom, 512–48.*, Public Law **Autumn** (2005), pp. 512–548.

[6] Carel, H. and I. Kidd, *Epistemic injustice in healthcare: a philosophical analysis*, Medicine, Health Care and Philosophy **17** (2014), p. 529–540.

[7] Chellas, B. F., "The logical form of imperatives," Stanford University, 1969.

[8] Chellas, B. F., "Modal logic: an introduction," Cambridge university press, 1980.

[9] Clark, H. H., "Using Language," Cambridge University Press, Cambridge, 1996.

[10] Dam, C. v., *Ongekend onrecht - verslag van de parlementaire ondervraging kinderopvangtoeslag*, Tweede Kamer der Staten Generaal (2020).

[11] Dastani, M., A. Herzig, J. Hulstijn and L. van der Torre, *Inferring trust*, in:

J. Leite, editor, *Proceedings of Fifth Workshop on Computational Logic in Multi-agent Systems (CLIMA V)* (2004), pp. 144–160.

[12] De Bruin, B., *Epistemic virtues in business*, Journal of Business Ethics **113** (2013), p. 583–595.

[13] de Bruin, B., *Epistemic injustice in finance*, Topoi **40** (2021), p. 755–763.

[14] Dembroff, R. and D. Whitcomb, *Content-focused epistemic injustice*, in: T. S. Gendler, J. Hawthorne and J. Chung, editors, *Oxford Studies in Epistemology - Volume 7*, Oxford University Press, 2022 pp. 48–70.

[15] Fricker, M., "Epistemic injustice." Oxford University Press, Oxford, 2007.

[16] Fricker, M., *Epistemic justice as a condition of political freedom?*, Synthese **190** (2013), pp. 1317–1332.

[17] Goldman, A. and C. O'Connor, *Social epistemology*, Stanford Encyclopedia of Philosophy (2021).

[18] Green, B. and A. Kak, *The false comfort of human oversight as an antidote to A.I. harm*, Slate **June** (2021).

[19] Holmes, C., *Report of the royal commission into the robodebt scheme*, Report, Commonwealth of Australia (2023).

[20] Hookway, C., *Some varieties of epistemic injustice: Reflections on fricker*, Episteme **7** (2010), p. 151–163.

[21] Icard, T., E. Pacuit and Y. Shoham, *Joint revision of beliefs and intention.*, in: *KR*, 2010, pp. 572–574.

[22] Kidd, I. J., J. Medina and G. Pohlhaus Jr., editors, "Handbook of Epistemic Injustice," Routledge, London, 2017, 1 edition.

[23] Parlementaire enquetecommissie fraudebeleid en dienstverlening, *Blind voor mens en recht*, Tweede Kamer der Staten Generaal (2024).

[24] Roehl, U. B., *Automated decision-making and good administration*, Government Information Quarterly **40** (2023), p. 101864.
URL   https://www.sciencedirect.com/science/article/pii/S0740624X23000643

[25] Searle, J. R., "Speech acts: an Essay in the Philosophy of Language," Cambridge University Press, Cambridge, 1969.

[26] Sheikh, H., C. Prins and E. Schrijvers, "Mission AI," Netherlands Council for Government Policy (WRR), Springer, The Hague, 2023.

[27] van Der Hoek, W., W. Jamroga and M. Wooldridge, *Towards a theory of intention revision*, Synthese **155** (2007), pp. 265–290.

Decidability of Horn Sequents over Intuitionistic Tense Logic S4

Zhe Yu [1] †

Department of Philosophy, Sun Yat-sen University
Guangzhou, China

Yiheng Wang [2] *†

Department of Philosophy, Sun Yat-sen University
Guangzhou, China

Zhe Lin [3] *

Department of Philosophy, Xiamen University
Xiamen, China

Abstract

In this paper, we investigate the decidability of Horn sequent in intuitionistic tense logic S4. By constructing the finite model property (FMP), the decidability result follows. This contributes to solving the open problem of decidability of intuitionistic tense logic.

Keywords: decidability, intuitionistic tense logic, Horn sequent.

1 Introduction

Ewald introduced intuitionistic tense logic in 1986 [9]. Within this framework, Ewald considered four tense operators, comprising two pairs of adjoint modalities: $(F(\Diamond), H(\blacksquare))$ and $(P(\blacklozenge), G(\square))$, which are integrated within intuitionistic logic. This foundational logic, denoted as IKt, was further explored by Figallo and Pelaitay in [10]. They provided an algebraic axiomatization of IKt-algebras, showing its soundness and completeness within the class of IKt-algebras. The intuitionistic tense S4 logic IS4t is clearly S4 extension of IKt.

---

[1] Email address: yuzh28@mail.sysu.edu.cn

[2] Email address: ianwang747@gmail.com

[3] Email address: pennyshaq@163.com

[4] † These authors contributed equally to this work. * Corresponding authors.

Intuitionistic logics and their modal expansions (cf. [26], [27], [29]), have attracted significant attention in artificial intelligence and computer science (cf. [6]). Intuitionistic tense or temporal logic was regarded as a fundamental component of temporal equilibrium logic (TEL), and important for answer-set semantics or answer-set programming (ASP), which are central in practical knowledge representation (cf. [11], [12], [19], [21],[23]). These semantic frameworks form the backbone for the DLV solver (cf. [15]), renowned for its implementation in disjunctive logic programming (DLP). Furthermore, intuitionistic temporal logic contributes significantly to the logical characterization of safety and liveness properties (cf. [18]).

The decidability problems of intuitionistic tense logic and its extensions are long-time open problems. In [9], Ewald initially demonstrated the decidability of IKt through the finite model property. However, Simpson later refuted this result in [27]. Various attempts have been made to deal with this gap. Several weakenings of intuitionistic tense logic are proven to be decidable. The intuitionistic modal logic given by Simpson can be conservatively extended to Ewald's system, and some results on intuitionistic tense logic are found in the literature. For example, some extensions of IK shown in Figure 7-5 [27] are decidable. The decidability of extensions like IK4, and IKD4 are still unknown. De Paiva showed a weaker version of IS4 called CS4 which doesn't satisfy the distributivity of $\Diamond$ over $\lor$ i.e. $\Diamond(\alpha \lor \beta) \Rightarrow \Diamond\alpha \lor \Diamond\beta$ is decidable [7]. Recently, Lutz Straßburger proved that IS4 is decidable [13], but this result can not be adapted to the tense version directly. The well-known intuitionistic modal logic MIPC, has been established as decidable in [3]. For intuitionistic tense logics, [16] showed that the disjunction-free fragment of IKt exhibits the FMP, thereby rendering it decidable. [17,24] provided another kind of weakening on dual modalities is decidable. [28] showed the cut elimination of this weak variant.

Horn clauses have been considered the basic content and provide many useful properties in computer science and artificial intelligence. These clauses are logical formulas of a particular rule-like form. The Horn clause discussed in [4] has the following form:

$$\alpha_1 \land \cdots \land \alpha_k \land \neg\alpha_{k+1} \land \cdots \land \neg\alpha_n \to \beta_1 \lor \cdots \lor \beta_m$$

where α_i $(1 \leq i \leq n)$ and β_j $(1 \leq j \leq m)$ are atomic formulas. One may consider the modal extension of such formulas. This can be possibly achieved under the language of intuitionistic tense logic with such tense or temporal operators. Inspired by the modal Horn clause defined in [5], we extend Horn classes to intuitionistic tense cases. An intuitionistic tense Horn clause is

$$\alpha_1 \lor \cdots \lor \alpha_n \to \beta_1 \lor \cdots \lor \beta_m$$

where α_i $(1 \leq i \leq n)$ and β_j $(1 \leq j \leq m)$ are formulas under connectives $\neg, \to, \land, \blacksquare, \square, \Diamond, \blacklozenge$. An intuitionistic tense Horn clause program is a finite set of such clauses [11].

In this paper, we consider the decidability of intuitionistic tense Horn sequent (cf. Section 3) over IS4t. IS4t is one of the most important and widely studied intuitionistic tense logic in literature. Moreover, the decidability of the general theory of IS4t remains unknown. Further, according to applications, one natural and interesting question is whether intuitionistic tense Horn sequent is decidable in IS4t. We show that some restricted sequents (formulas) are decidable in IS4t by proving finite model property, which yields the decidability of intuitionistic tense Horn sequent of IS4t. Surely this result only provides a partial answer to the open problem but this is still a beneficial supplement. The study of the finite model property (FMP) for intuitionistic modal logic has a long tradition. H.Ono [22] and G.S.Fischer [25] showed the FMP of the well-known MIPC and some results over MIPC were also reached [1]. F.Wolter and M.Zakhariaschev [29] showed that some K4 extensions of a basic intuitionistic tense logic, i.e. $\Diamond, \blacklozenge, \Box, \blacksquare$ satisfying the relations: $\alpha \to \Box\blacklozenge\alpha$ and $\alpha \to \blacksquare\Diamond\alpha$ have FMP and thus decidable. W.Dzik, J.Järvinen, and M.Kondo addressed the problem of FMP of intuitionistic propositional logic with Galois connections [8]. P.Balbiani and M.Diéguez [2] proved that both the temporal here and there (THT) and its intuitionistic variants (ITLp) have FMP.

This paper is organized as follows. Section 2 gives some preliminaries of algebra and sequent calculus with soundness and completeness proved. Section 3 proves the FMP about the restricted set of formulas i.e. the so-called Horn sequents. Section 4 gives some concluding remarks.

2 Algebra and Calculus

In this section, we give some algebraic and logical preliminaries.

Definition 1 ([10]) An *intuitionistic tense algebra* (IKt-algebra for short) is a structure $\mathcal{A} = (A, \wedge, \vee, \to, \Diamond, \Box, \blacklozenge, \blacksquare, 0, 1)$, where $\mathcal{A} = (A, \wedge, \vee, \to, 0, 1)$ is a Heyting algebra, and $\Diamond, \Box, \blacklozenge, \blacksquare$ are unary operators on A satisfying the following conditions: for all $a, b \in A$,

(t1) $\Box 1 = 1$ and $\blacksquare 1 = 1$.
(t2) $\Box(a \wedge b) = \Box a \wedge \Box b$ and $\blacksquare(a \wedge b) = \blacksquare a \wedge \blacksquare b$.
(t3) $a \le \blacksquare\Diamond a$ and $a \le \Box\blacklozenge a$.
(t4) $\Diamond 0 = 0$ and $\blacklozenge 0 = 0$.
(t5) $\Diamond(a \vee b) = \Diamond a \vee \Diamond b$ and $\blacklozenge(a \vee b) = \blacklozenge a \vee \blacklozenge b$.
(t6) $\Diamond\blacksquare a \le a$ and $\blacklozenge\Box a \le a$.
(t7) $\Diamond(a \to b) \le \Box a \to \Diamond b$ and $\blacklozenge(a \to b) \le \blacksquare a \to \blacklozenge b$.

An IS4t-algebra is obtained from IKt-algebra by adding $\sharp a \le \sharp\sharp a$ and $\sharp a \le a$ where $\sharp \in \{\Box, \blacksquare\}$.

Definition 2 The set of all formulas $\mathcal{F}$ is defined inductively as follows:

$$\mathcal{F} \ni \alpha ::= p \mid \top \mid \bot \mid (\alpha_1 \to \alpha_2) \mid (\alpha_1 \wedge \alpha_2) \mid (\alpha_1 \vee \alpha_2) \mid \Diamond\alpha \mid \Box\alpha \mid \blacklozenge\alpha \mid \blacksquare\alpha$$

where $p \in \mathbf{Var} = \{p_i : i < \omega\}$, a denumerable set of variables. We use the abbreviation $\neg\alpha := \alpha \to \bot$. Formulas belong to $\mathbf{Atom} = \mathbf{Var} \cup \{\bot, \top\}$ are called *atomic*. The *complexity* of a formula α, denoted by $c(\alpha)$, is defined as usual. Let $Sub(\alpha)$ be the set of all subformulas of α. For a set of formulas Γ, let $Sub(\Gamma) = \bigcup_{\alpha \in \Gamma} Sub(\alpha)$. A *substitution* is a homomorphism $\sigma : \mathcal{F} \to \mathcal{F}$.

Definition 3 Let the comma, $\circ$ and $\bullet$ be structural counterparts for $\wedge$, $\diamond$ and $\blacklozenge$ respectively. The set of all formula structures $\mathcal{FS}$ is defined inductively as follows:

$$\mathcal{FS} \ni \Gamma ::= \alpha \mid (\Gamma_1, \Gamma_2) \mid \circ\Gamma \mid \bullet\Gamma$$

Let $\mathcal{FS}^\epsilon = \mathcal{FS} \cup \{\epsilon\}$ where ϵ stands for the empty formula structure. By $f(\Gamma)$ we denote the formula obtained from Γ by replacing all structure operators with their corresponding formula connectives. A *sequent* is an expression of the form $\Gamma \Rightarrow \beta$ where $\Gamma \in \mathcal{FS}^\epsilon$ is a formula structure and $\beta \in \mathcal{F}$ is a formula.

Definition 4 Let $-$ be the symbol called the *position*. A *context* $\Gamma[-]$ is a formula structure $\Gamma \in \mathcal{FS}^\epsilon$ together with a designated position $[-]$ which can be filled with a formula structure. The set of all contexts $\mathcal{C}$ is defined inductively as follows:

$$\mathcal{C} \ni \Gamma[-] ::= - \mid (\Gamma_1[-], \Gamma_2) \mid (\Gamma_1, \Gamma_2[-]) \mid \circ\Gamma[-] \mid \bullet\Gamma[-]$$

Example 5 Let expression $\Gamma[-] = \circ^5 \bullet^{10} ((-, q), p \wedge q)$ be a context with position $-$. If we replace the formula structure $\Delta = \bullet(p, (q, \circ q))$ for the position $-$ in $\Gamma[-]$, we get the formula structure $\Gamma[\Delta] = \circ^5 \bullet^{10} ((\bullet(p, (q, \circ q)), q), p \wedge q)$.

Definition 6 The sequent calculus GIS4t consists of the following axiom schemata and sequent rules: for $i \in \{1, 2\}$ and $\star \in \{\circ, \bullet\}$,

(1) Axiom schemata:

$$\text{(Id)} \ \alpha \Rightarrow \alpha$$

(2) Logical rules:

$$\frac{\Gamma[\top] \Rightarrow \beta}{\Gamma[\Delta] \Rightarrow \beta}(\top) \quad \frac{\Delta \Rightarrow \bot}{\Gamma[\Delta] \Rightarrow \beta}(\bot) \quad \frac{\Gamma[\alpha_1, \alpha_2] \Rightarrow \beta}{\Gamma[\alpha_1 \wedge \alpha_2] \Rightarrow \beta}(\wedge L) \quad \frac{\Gamma \Rightarrow \beta_1 \quad \Gamma \Rightarrow \beta_2}{\Gamma \Rightarrow \beta_1 \wedge \beta_2}(\wedge R)$$

$$\frac{\Gamma[\alpha_1] \Rightarrow \beta \quad \Gamma[\alpha_2] \Rightarrow \beta}{\Gamma[\alpha_1 \vee \alpha_2] \Rightarrow \beta}(\vee L) \quad \frac{\Gamma \Rightarrow \beta_i}{\Gamma \Rightarrow \beta_1 \vee \beta_2}(\vee R)$$

$$\frac{\Delta \Rightarrow \alpha_1 \quad \Gamma[\alpha_2] \Rightarrow \beta}{\Gamma[\Delta, \alpha_1 \to \alpha_2] \Rightarrow \beta}(\to L) \quad \frac{\beta_1, \Gamma \Rightarrow \beta_2}{\Gamma \Rightarrow \beta_1 \to \beta_2}(\to R)$$

(3) Tense rules:

$$\frac{\Gamma[\circ\alpha] \Rightarrow \beta}{\Gamma[\diamond\alpha] \Rightarrow \beta}(\diamond L) \quad \frac{\Gamma \Rightarrow \beta}{\circ\Gamma \Rightarrow \diamond\beta}(\diamond R) \quad \frac{\Gamma[\bullet\alpha] \Rightarrow \beta}{\Gamma[\blacklozenge\alpha] \Rightarrow \beta}(\blacklozenge L) \quad \frac{\Gamma \Rightarrow \beta}{\bullet\Gamma \Rightarrow \blacklozenge\beta}(\blacklozenge R)$$

$$\frac{\Gamma[\alpha] \Rightarrow \beta}{\Gamma[\circ\blacksquare\alpha] \Rightarrow \beta}(\blacksquare L) \quad \frac{\circ\Gamma \Rightarrow \beta}{\Gamma \Rightarrow \blacksquare\beta}(\blacksquare R) \quad \frac{\Gamma[\alpha] \Rightarrow \beta}{\Gamma[\bullet\Box\alpha] \Rightarrow \beta}(\Box L) \quad \frac{\bullet\Gamma \Rightarrow \beta}{\Gamma \Rightarrow \Box\beta}(\Box R)$$

(4) Structural rules:

$$\frac{\Gamma[\circ\Delta_1, \circ\Delta_2] \Rightarrow \beta}{\Gamma[\circ(\Delta_1, \Delta_2)] \Rightarrow \beta}(\text{Con}_\circ) \qquad \frac{\Gamma[\bullet\Delta_1, \bullet\Delta_2] \Rightarrow \beta}{\Gamma[\bullet(\Delta_1, \Delta_2)] \Rightarrow \beta}(\text{Con}_\bullet)$$

$$\frac{\Gamma[\alpha, \alpha] \Rightarrow \beta}{\Gamma[\alpha] \Rightarrow \beta}(\text{Con}_F) \qquad \frac{\Gamma[\Delta_i] \Rightarrow \beta}{\Gamma[\Delta_1, \Delta_2] \Rightarrow \beta}(\text{Wk}) \qquad \frac{\Gamma[\Delta_1, \Delta_2] \Rightarrow \beta}{\Gamma[\Delta_2, \Delta_1] \Rightarrow \beta}(\text{Ex})$$

$$\frac{\Gamma[\circ(\bullet\Delta_1, \Delta_2)] \Rightarrow \beta}{\Gamma[\Delta_1, \circ\Delta_2] \Rightarrow \beta}(\text{K}_{\circ\bullet}) \qquad \frac{\Gamma[\bullet(\circ\Delta_1, \Delta_2)] \Rightarrow \beta}{\Gamma[\Delta_1, \bullet\Delta_2] \Rightarrow \beta}(\text{K}_{\bullet\circ})$$

$$\frac{\Gamma[\star\alpha] \Rightarrow \beta}{\Gamma[\alpha] \Rightarrow \beta}(\text{T}_\star) \qquad \frac{\Gamma[\star\alpha] \Rightarrow \beta}{\Gamma[\star\star\alpha] \Rightarrow \beta}(4_\star)$$

(5) Cut rule:

$$\frac{\Delta \Rightarrow \alpha \quad \Gamma[\alpha] \Rightarrow \beta}{\Gamma[\Delta] \Rightarrow \beta}(\text{Cut})$$

Fact 7 *The following sequent rules are admissible in* GIS4t*:*

$$\frac{\alpha_1 \Rightarrow \beta_1 \quad \alpha_2 \Rightarrow \beta_2}{\alpha_1 \wedge \alpha_2 \Rightarrow \beta_2 \wedge \beta_2}(\wedge) \qquad \frac{\alpha_1 \Rightarrow \beta_1 \quad \alpha_2 \Rightarrow \beta_2}{\alpha_1 \vee \alpha_2 \Rightarrow \beta_2 \vee \beta_2}(\vee)$$

$$\frac{\alpha \Rightarrow \alpha \quad \beta \Rightarrow \beta}{\alpha \to \beta \Rightarrow \alpha \to \beta}(\to) \qquad \frac{\Gamma[\Delta, \Delta] \Rightarrow \beta}{\Gamma[\Delta] \Rightarrow \beta}(\text{Con}) \qquad \frac{\alpha \Rightarrow \beta}{\flat\alpha \Rightarrow \flat\beta}(\text{Mon})(\flat \in \{\Diamond, \Box, \blacklozenge, \blacksquare\})$$

$$\frac{\Gamma[\Delta_1, (\Delta_2, \Delta_3)] \Rightarrow \beta}{\Gamma[(\Delta_1, \Delta_2), \Delta_3] \Rightarrow \beta}(\text{As}_1) \qquad \frac{\Gamma[(\Delta_1, \Delta_2), \Delta_3] \Rightarrow \beta}{\Gamma[\Delta_1, (\Delta_2, \Delta_3)] \Rightarrow \beta}(\text{As}_2)$$

Proposition 8 *The following sequents are derivable in* GIS4t*:*

(1) $\Box\top \Leftrightarrow \top$ *and* $\blacksquare\top \Leftrightarrow \top$.
(2) $\Box(\alpha \wedge \beta) \Leftrightarrow \Box\alpha \wedge \Box\beta$ *and* $\blacksquare(\alpha \wedge \beta) \Leftrightarrow \blacksquare\alpha \wedge \blacksquare\beta$.
(3) $\alpha \Rightarrow \blacksquare\Diamond\alpha$ *and* $\alpha \Rightarrow \Box\blacklozenge\alpha$.
(4) $\Diamond\bot \Leftrightarrow \bot$ *and* $\blacklozenge\bot \Leftrightarrow \bot$.
(5) $\Diamond(\alpha \vee \beta) \Leftrightarrow \Diamond\alpha \vee \Diamond\beta$ *and* $\blacklozenge(\alpha \vee \beta) \Leftrightarrow \blacklozenge\alpha \vee \blacklozenge\beta$.
(6) $\Diamond\blacksquare\alpha \Rightarrow \alpha$ *and* $\blacklozenge\Box\alpha \Rightarrow \alpha$.
(7) $\Diamond(\alpha \to \beta) \Rightarrow \Box\alpha \to \Diamond\beta$ *and* $\blacklozenge(\alpha \to \beta) \Rightarrow \blacksquare\alpha \to \blacklozenge\beta$.
(8) $\Box\alpha \Rightarrow \Box\Box\alpha$ *and* $\blacksquare\alpha \Rightarrow \blacksquare\blacksquare\alpha$.
(9) $\Box\alpha \Rightarrow \alpha$ *and* $\blacksquare\alpha \Rightarrow \alpha$.

Proof. We only show the proofs of following cases:

(1)

$$\frac{\Box\top \Rightarrow \Box\top}{\top \Rightarrow \Box\top}(\top)$$

$$\frac{\top \Rightarrow \top}{\Box\top \Rightarrow \top}(\top)$$

(2)

$$\cfrac{\cfrac{\cfrac{\alpha \Rightarrow \alpha}{\alpha \wedge \beta \Rightarrow \alpha}\,(\wedge\text{L})}{\Box(\alpha \wedge \beta) \Rightarrow \Box\alpha}\,(\text{Mon}) \quad \cfrac{\cfrac{\beta \Rightarrow \beta}{\alpha \wedge \beta \Rightarrow \beta}\,(\wedge\text{L})}{\Box(\alpha \wedge \beta) \Rightarrow \Box\beta}\,(\text{Mon})}{\Box(\alpha \wedge \beta) \Rightarrow \Box\alpha \wedge \Box\beta}\,(\wedge\text{R})$$

$$\cfrac{\cfrac{\cfrac{\cfrac{\cfrac{\alpha \Rightarrow \alpha}{\bullet\Box\alpha \Rightarrow \alpha}\,(\Box\text{L})}{\bullet\Box\alpha, \bullet\Box\beta \Rightarrow \alpha}\,(\text{Wk}) \quad \cfrac{\cfrac{\beta \Rightarrow \beta}{\bullet\Box\beta \Rightarrow \beta}\,(\Box\text{L})}{\bullet\Box\alpha, \bullet\Box\beta \Rightarrow \beta}\,(\text{Wk})}{\bullet\Box\alpha, \bullet\Box\beta \Rightarrow \alpha \wedge \beta}\,(\wedge\text{R})}{\bullet(\Box\alpha, \Box\beta) \Rightarrow \alpha \wedge \beta}\,(\text{Con}_\bullet)}{\cfrac{\Box\alpha, \Box\beta \Rightarrow \Box(\alpha \wedge \beta)}{\Box\alpha \wedge \Box\beta \Rightarrow \Box(\alpha \wedge \beta)}\,(\wedge\text{R})}\,(\Box\text{R})$$

(3)

$$\cfrac{\cfrac{\alpha \Rightarrow \alpha}{\circ\alpha \Rightarrow \Diamond\alpha}\,(\text{Mon})}{\alpha \Rightarrow \blacksquare\Diamond\alpha}\,(\blacksquare\text{R})$$

(4)

$$\cfrac{\cfrac{\bot \Rightarrow \bot}{\circ\bot \Rightarrow \bot}\,(\bot)}{\Diamond\bot \Rightarrow \bot}\,(\Diamond\text{L})$$

$$\cfrac{\bot \Rightarrow \bot}{\bot \Rightarrow \Diamond\bot}\,(\bot)$$

(5)

$$\cfrac{\cfrac{\cfrac{\cfrac{\alpha \Rightarrow \alpha}{\circ\alpha \Rightarrow \Diamond\alpha}\,(\Diamond\text{R})}{\circ\alpha \Rightarrow \Diamond\alpha \vee \Diamond\beta}\,(\vee\text{R}) \quad \cfrac{\cfrac{\beta \Rightarrow \beta}{\circ\beta \Rightarrow \Diamond\beta}\,(\Diamond\text{R})}{\circ\beta \Rightarrow \Diamond\alpha \vee \Diamond\beta}\,(\vee\text{R})}{\circ(\alpha \vee \beta) \Rightarrow \Diamond\alpha \vee \Diamond\beta}\,(\vee\text{L})}{\Diamond(\alpha \vee \beta) \Rightarrow \Diamond\alpha \vee \Diamond\beta}\,(\Diamond\text{L})$$

$$\cfrac{\cfrac{\cfrac{\cfrac{\alpha \Rightarrow \alpha}{\alpha \Rightarrow \alpha \vee \beta}\,(\vee\text{R})}{\circ\alpha \Rightarrow \Diamond(\alpha \vee \beta)}\,(\Diamond\text{R})}{\Diamond\alpha \Rightarrow \Diamond(\alpha \vee \beta)}\,(\Diamond\text{L}) \quad \cfrac{\cfrac{\cfrac{\beta \Rightarrow \beta}{\beta \Rightarrow \alpha \vee \beta}\,(\vee\text{R})}{\circ\beta \Rightarrow \Diamond(\alpha \vee \beta)}\,(\Diamond\text{R})}{\Diamond\beta \Rightarrow \Diamond(\alpha \vee \beta)}\,(\Diamond\text{L})}{\Diamond\alpha \vee \Diamond\beta \Rightarrow \Diamond(\alpha \vee \beta)}\,(\vee\text{L})$$

(6)

$$\cfrac{\cfrac{\alpha \Rightarrow \alpha}{\circ\blacksquare\alpha \Rightarrow \alpha}\,(\blacksquare\text{L})}{\Diamond\blacksquare\alpha \Rightarrow \alpha}\,(\Diamond\text{L})$$

(7)

$$\dfrac{\dfrac{\alpha \Rightarrow \alpha \quad \beta \Rightarrow \beta}{\alpha, \alpha \to \beta \Rightarrow \beta} (\to\text{L})}{\dfrac{\circ(\alpha, \alpha \to \beta) \Rightarrow \Diamond\beta}{\dfrac{\circ(\bullet\Box\alpha, \alpha \to \beta) \Rightarrow \Diamond\beta}{\dfrac{\Box\alpha, \circ(\alpha \to \beta) \Rightarrow \Diamond\beta}{\dfrac{\Box\alpha, \Diamond(\alpha \to \beta) \Rightarrow \Diamond\beta}{\Diamond(\alpha \to \beta) \Rightarrow \Box\alpha \to \Diamond\beta} (\to\text{L})} (\Diamond\text{L})} (\text{K}_{\circ\bullet})} (\Box\text{L})} (\Diamond\text{R})}$$

(8)

$$\dfrac{\dfrac{\dfrac{\alpha \Rightarrow \alpha}{\bullet\Box\alpha \Rightarrow \alpha} (\Box\text{L})}{\bullet\bullet\Box\alpha \Rightarrow \alpha} (4_\star)}{\Box\alpha \Rightarrow \Box\Box\alpha} (\Box\text{R})$$

(9)

$$\dfrac{\dfrac{\alpha \Rightarrow \alpha}{\bullet\Box\alpha \Rightarrow \alpha} (\Box\text{L})}{\Box\alpha \Rightarrow \alpha} (\text{T}_\star)$$

$\square$

Theorem 9 GIS4t *are sound and complete w.r.t.* **IS4t**, *the variety of IS4ts.*

Proof. The soundness proceeds by the induction on the height of derivation. For completeness result, it suffices to show that for any sequent $\Gamma \Rightarrow \beta$, if $\nvdash \Gamma \Rightarrow \beta$, then $\nvDash \Gamma \Rightarrow \beta$. It can be proved by standard construction. Let $[\![\alpha]\!] = \{\beta | \vdash \alpha \Leftrightarrow \beta\}$. Let A be the set of all $[\![\alpha]\!]$. One defines $\{\wedge', \vee', \to', P, F, \top', \bot'\}$ on A as follows:

$$[\![\alpha_1]\!] \wedge' [\![\alpha_2]\!] = [\![\alpha_1 \wedge \alpha_2]\!] \quad [\![\alpha_1]\!] \vee' [\![\alpha_2]\!] = [\![\alpha_1 \vee \alpha_2]\!] \quad [\![\alpha_1]\!] \to' [\![\alpha_2]\!] = [\![\alpha_1 \to \alpha_2]\!]$$

$$F[\![\alpha]\!] = [\![\Diamond\alpha]\!] \quad P[\![\alpha]\!] = [\![\blacksquare\alpha]\!] \quad \top' = [\![\top]\!] \quad \bot' = [\![\bot]\!]$$

The order is defined as $[\![\alpha_1]\!] \leq' [\![\alpha_2]\!]$ iff $[\![\alpha_1]\!] \wedge' [\![\alpha_2]\!] = [\![\alpha_1]\!]$. Thus $[\![\alpha_1]\!] \leq' [\![\alpha_2]\!]$ iff $\vdash \alpha_1 \Rightarrow \alpha_2$. Define an assignment $\sigma : \mathbf{Var} \to A$ such that $\sigma(p) = [p]$. By induction on the complexity of the formula, one shows that $\widehat{\sigma}(\alpha) = [\![\alpha]\!]$ for any formula α. Suppose that $\vDash \Gamma \Rightarrow \beta$. Then $\widehat{\sigma}(\Gamma) \leq \widehat{\sigma}(\beta)$. Hence $\vdash \Gamma \Rightarrow \beta$, which yields a contradiction. This completes the proof. $\square$

Lemma 10 *The following hold in cut-free* GIS4t*:*

(1) *If* $\vdash \Gamma[\alpha_1 \wedge \alpha_2] \Rightarrow \beta$, *then* $\vdash \Gamma[\alpha_1, \alpha_2] \Rightarrow \beta$.
(2) *If* $\vdash \Gamma[\alpha_1 \vee \alpha_2] \Rightarrow \beta$, *then* $\vdash \Gamma[\alpha_1] \Rightarrow \beta$ *and* $\vdash \Gamma[\alpha_2] \Rightarrow \beta$.
(3) *If* $\vdash \Gamma[\Diamond\alpha] \Rightarrow \beta$, *then* $\vdash \Gamma[\circ\alpha] \Rightarrow \beta$.
(4) *If* $\vdash \Gamma[\blacklozenge\alpha] \Rightarrow \beta$, *then* $\vdash \Gamma[\bullet\alpha] \Rightarrow \beta$.

Proof. The proof proceeds by the induction on the height of derivation n. We only show the proof of (1). Assume $\vdash \Gamma[\alpha_1 \wedge \alpha_2] \Rightarrow \beta$. Let $n = 0$, i.e., $\alpha_1 \wedge \alpha_2 \Rightarrow \alpha_1 \wedge \alpha_2$. Clearly, one has $\vdash \alpha_1, \alpha_2 \Rightarrow \alpha_1 \wedge \alpha_2$. Assume $n > 0$.

If $\alpha_1 \wedge \alpha_2$ is principal in (R), then (R) is ($\wedge$L) or ($\top$) or ($\bot$). For ($\wedge$L), one has $\vdash^n \Gamma[\alpha_1, \alpha_2] \Rightarrow p$ simply by the premise. For other cases, one has $\vdash^n \Gamma[\alpha_1, \alpha_2] \Rightarrow p$ by (R) on its premise. If $\alpha_1 \wedge \alpha_2$ is not principal in (R), then one has $\vdash^n \Gamma[\alpha_1, \alpha_2] \Rightarrow p$ by induction hypothesis and rule (R). □

Theorem 11 (Cut-elimination) GIS4t *enjoys the cut elimination.*

Proof. The proof can be checked in [16,17,28], which shows the cut elimination of weak intuitionistic tense logic. The proof method is quite similar. □

Corollary 12 GIS4t *enjoys the subformula property.*

3   Decidability of Horn Sequent

In this section, we consider a set of restricted sequents $\mathcal{RS}$ i.e. so-called Horn sequents which are constituted by formulas where the disjunctions are not included in the implications or box (including $\square$ and $\blacksquare$) connectives. For instance, sequents like $\blacksquare((\alpha \wedge \beta) \vee \gamma) \Rightarrow \gamma$, $\alpha \vee \beta \rightarrow \gamma \Rightarrow \gamma \vee \alpha$ don't belong to such a set. On the contrary, general Horn clauses are certainly examples of such sequents. We will show that for any sequents in $\mathcal{RS}$, if it is not derivable in GIS4t, then there exists a finite IS4t algebra refuting it. Thus by [14], all kinds of such sequents in $\mathcal{RS}$ and Horn sequent are decidable.

We now construct some special sets of formulas. Let T be a set of formulas satisfying that disjunctions are not included in the implications or box connectives and closed under subformulas. Let $T^{\#}$ be the $\wedge, \rightarrow$ closure of T and $\overline{T}$ be the $\wedge, \vee$ closure of $T^{\#}$. Clearly there is no formulas containing $\vee$ included in $\rightarrow$ or $\blacksquare, \square$. Moreover, if T is finite then $T^{\#}$ is finite up to equivalence relations $\Leftrightarrow$. Hence $\overline{T}$ is finite up to equivalence.

Given a sequent $\Gamma[[\Delta_1[\Sigma_{11}]\ldots[\Sigma_{1m_1}]]\ldots[\Delta_n[\Sigma_{n1}]\ldots[\Sigma_{nm_n}]]] \Rightarrow \beta$ satisfying that there is no formulas of the form $\alpha_1 \vee \alpha_2$ contained in any $\Sigma_{ij} \Rightarrow \theta_{ij}$ where $1 \leq i \leq n, 1 \leq j \leq max(m_1, \ldots, m_n)$. Let T containing $\bot, \top$ be the set of all subformulas of formulas in this sequents.

Lemma 13 *If* $\vdash \Gamma[[\Delta_1[\Sigma_{11}]\ldots[\Sigma_{1m_1}]]\ldots[\Delta_n[\Sigma_{n1}]\ldots[\Sigma_{nm_n}]]] \Rightarrow \beta$, *then there are* $\gamma_i, \theta_{ij} \in T^{\#}$ *such that* $\vdash \Sigma_{ij} \Rightarrow \theta_{ij}$ *where* $1 \leq i \leq n, 1 \leq j \leq max(m_1, \ldots, m_n)$, $\vdash \Delta_i[\theta_{i1}]\ldots[\theta_{im_i}] \Rightarrow \gamma_i$, *and* $\vdash \Gamma[\gamma_1]\ldots[\gamma_i] \Rightarrow \beta$.

Proof. We proceed by induction on the length of the cut-free proof. Suppose that the sequent is ended by rule (R). One suffices to show that any structure considered here has an interpolant γ. Thus without loss of generality, we only display one structure Δ. Assume that $\vdash \Gamma'[\Delta] \Rightarrow \beta$ and ended by rule (R). If (R) is the right rule, then the claim is held by induction hypothesis and (R). Let (R) be a left rule. For instance, let (R) be $(K_{\circ\bullet})$ (Other cases can be treated similarly). Assume that the premise is $\Gamma''[\circ(\Sigma_1, \bullet\Sigma_2)] \Rightarrow \alpha$ and the conclusion is $\Gamma''[\Sigma_1, \circ\Sigma_2] \Rightarrow \alpha$. Let $\Delta = \circ\Sigma_2$. Then by induction hypothesis, one obtains $\vdash \Sigma_1 \Rightarrow \gamma_1, \vdash \circ(\gamma_2, \bullet\Sigma_2) \Rightarrow \gamma_1$ and $\vdash \Gamma''[\gamma_1] \Rightarrow \alpha$ where $\gamma_1, \gamma_2 \in T^{\#}$. Then by $(K_{\circ\bullet})$ and ($\rightarrow$R), one gets $\vdash \bullet\Sigma_2 \Rightarrow \gamma_1 \rightarrow \gamma_2$. Moreover by ($\rightarrow$L), one gets $\vdash \Gamma''[\gamma_1, \gamma_1 \rightarrow \gamma_2] \Rightarrow \alpha$. Hence $\vdash \Gamma''[\Sigma_1, \gamma_1 \rightarrow \gamma_2] \Rightarrow \alpha$. Since $\gamma_1 \rightarrow \gamma_2 \in T^{\#}$. Thus $\gamma_1 \rightarrow \gamma_2$ is the required interpolant. Let (R) be ($\vee$L).

Clearly the principal formula $\alpha_1 \vee \alpha_2$ can not contained in Δ. Assume that the premises are $\vdash \Gamma''[\alpha_1][\Delta] \Rightarrow \alpha$ and $\vdash \Gamma''[\alpha_2][\Delta] \Rightarrow \alpha$. Then by induction hypothesis there are γ_1, γ_2 such that $\vdash \Delta \Rightarrow \gamma_j$ and $\vdash \Gamma''[\alpha_i][\gamma_j] \Rightarrow \alpha$ where $i, j \in \{1, 2\}$. Thus $\vdash \Delta \Rightarrow \gamma_1 \wedge \gamma_2$ and $\vdash \Gamma''[\alpha_1 \vee \alpha_2][\gamma_1 \wedge \gamma_2] \Rightarrow \alpha$. Since $\gamma_1 \wedge \gamma_2 \in T^\#$, it is the desired interpolant. □

Definition 14 [Order on $\mathrm{sf}(\overline{T})$] Define $\leq$ on $\mathrm{sf}(\overline{T})$ as follows: $\Delta_1 \leq \Delta_2$ iff for any $\alpha \in \overline{T}$, if $\vdash_{\mathsf{GIS4t}} \Gamma[\Delta_2] \Rightarrow \alpha$, then $\vdash_{\mathsf{GIS4t}} \Gamma[\Delta_1] \Rightarrow \alpha$ where $\Gamma[\Delta_1], \Gamma[\Delta_2] \in \mathrm{sf}(\overline{T})$ and $\alpha \in \overline{T}$.

Let $\Delta_1 \sim \Delta_2$ be $\Delta_1 \leq \Delta_2$ and $\Delta_2 \sim \Delta_1$, then clearly $\sim$ is an equivalence relation. Let $[\![\alpha]\!] := \{\Delta \in \mathrm{sf}(\mathrm{sf}(\overline{T})) | \alpha \sim \Delta\}$ for any $\alpha \in \mathrm{sf}(\overline{T})$.

Lemma 15 *Let T be a finite set containing $\top, \bot$ and closed under subformula. Define $|T^\#| = \{[\![\alpha]\!] : \alpha \in T^\#\}$ and $|\overline{T}| = \{[\![\alpha]\!] : \alpha \in \overline{T}\}$, then both $|T|$ and $|T^\#|$ are finite.*

Proof. Since T is finite, $|T^\#| = \{[\![\alpha]\!] : \alpha \in T^\#\}$ is finite. [20] showed that the variety of implicative semi-lattice is locally finite. Thus $\mathcal{K}$ generated by $|T^\#|$ is finite. Because the domain of $\mathcal{K}$ is $|T^\#| = \{[\![\alpha]\!] : \alpha \in T^\#\}$, hence $|T^\#|$ is finite. Therefore, $|\overline{T}|$ is finite by construction. □

For the rest of the paper, we let $T^\#$ and $\overline{T}$ be constructed from a finite formula set.

Lemma 16 *For any $\alpha \in T^\#$, there is $\beta \in T^\#$ s.t. $\star\alpha \sim \beta$ where $\star \in \{\circ, \bullet\}$.*

Proof. Suppose that $\vdash \Gamma[\star\alpha] \Rightarrow \theta$. Then by Lemma 13, there is a $\gamma \in T^\#$ such that $\vdash \star\alpha \Rightarrow \gamma$ and $\vdash \Gamma[\gamma] \Rightarrow \theta$. Since $|T^\#|$ is finite, let $\{\gamma_1, \ldots, \gamma_n\}$ be the set of all interpolants for all $\Gamma'[-]$ and θ such that $\vdash \Gamma[\star\alpha] \Rightarrow \theta$. Let $\beta = \gamma_1 \wedge \ldots \wedge \gamma_n$. Then $\beta \in T^\#$. Clearly by $(\wedge \mathrm{R})$ and $(\wedge \mathrm{L})$, one has (1) $\vdash \star\alpha \Rightarrow \beta$ and (2) $\vdash \Gamma[\beta] \Rightarrow \theta$. From (2) one has $\beta \leq \star\alpha$. Suppose that $\vdash \Gamma[\beta] \Rightarrow \pi$ for any π. Applying (Cut) to (1) and the assumption, one has $\vdash \Gamma[\star\alpha] \Rightarrow \pi$. Hence $\star\alpha \leq \beta$. □

Corollary 17 *For any $\alpha \in \overline{T}$, there is $\beta \in \overline{T}$ s.t. $\star\alpha \sim \beta$ where $\star \in \{\circ, \bullet\}$.*

Definition 18 [Quotient algebra] The quotient algebra of $\overline{T}$ is a structure $\mathcal{Q} = (|\overline{T}|, f_\wedge, f_\vee, f_\to, f_\diamond, f_\blacklozenge, f_\square, f_\blacksquare, f_0, f_1)$, where: for any $[\![\alpha]\!], [\![\beta]\!] \in |\overline{T}|$,

(i) $[\![\alpha]\!] f_\wedge [\![\beta]\!] = [\![\alpha \wedge \beta]\!]$.

(ii) $[\![\alpha]\!] f_\vee [\![\beta]\!] = [\![\alpha \vee \beta]\!]$.

(iii) $[\![\alpha]\!] f_\to [\![\beta]\!] = [\![\gamma_1 \vee \ldots \vee \gamma_n]\!]$ where $\gamma_i \in \overline{T}$ s.t. $[\![\alpha]\!] f_\wedge [\![\gamma_i]\!] \leq [\![\beta]\!]$ for any $i \in \{1, \ldots n\}$.

(iv) $f_0 = [\![\bot]\!]$.

(v) $f_1 = [\![\top]\!]$.

(vi) $f_\diamond [\![\alpha]\!] = [\![\gamma]\!]$ where $\gamma \in \overline{T}$ s.t. $\gamma \sim \circ\alpha$.

(vii) $f_\blacklozenge [\![\alpha]\!] = [\![\gamma]\!]$ where $\gamma \in \overline{T}$ s.t. $\gamma \sim \bullet\alpha$.

(viii) $f_\square [\![\alpha]\!] = [\![\gamma_1 \vee \ldots \vee \gamma_n]\!]$ where $\gamma_i \in \overline{T}$ s.t. $f_\blacklozenge [\![\gamma_i]\!] \leq [\![\alpha]\!]$ for any $i \in \{1, \ldots n\}$.

(ix) $f_\blacksquare [\![\alpha]\!] = [\![\gamma_1 \vee \ldots \vee \gamma_n]\!]$ where $\gamma_i \in \overline{T}$ s.t. $f_\diamond [\![\gamma_i]\!] \leq [\![\alpha]\!]$ for any $i \in \{1, \ldots n\}$.

One can easily check that $\mathcal{Q}$ is well-defined. The order of $\mathcal{Q}$ is defined by $f_\wedge$ as: $[\![\alpha]\!] \leq [\![\beta]\!]$ iff $[\![\alpha]\!] f_\wedge [\![\beta]\!] = [\![\alpha]\!]$.

Lemma 19 $\vdash \alpha \Rightarrow \beta$ iff $[\![\alpha]\!] \leq [\![\beta]\!]$ where $\alpha, \beta \in \overline{T}$.

Proof. Assume that $\vdash \alpha \Rightarrow \beta$. Then $\vdash \alpha \Leftrightarrow \alpha \wedge \beta$. Thus $[\![\alpha]\!] \leq [\![\beta]\!]$. Conversely let $[\![\alpha]\!] \leq [\![\beta]\!]$. Then $\alpha \wedge \beta \sim \alpha$. Clearly $\vdash \alpha \wedge \beta \Rightarrow \beta$. Hence $\vdash \alpha \Rightarrow \beta$. □

Lemma 20 *The following conditions hold for $\mathcal{Q}$: for any $[\![\alpha]\!], [\![\beta]\!]$ and $[\![\gamma]\!] \in |\overline{T}|$,*

(i) $f_\diamond([\![\alpha]\!] \vee [\![\beta]\!]) = f_\diamond[\![\alpha]\!] \vee f_\diamond[\![\beta]\!]$ *and* $f_\blacklozenge([\![\alpha]\!] \vee [\![\beta]\!]) = f_\blacklozenge[\![\alpha]\!] \vee f_\blacklozenge[\![\beta]\!]$.
(ii) $f_\diamond f_0 = f_0$ *and* $f_\blacklozenge f_0 = f_0$.
(iii) $f_\diamond f_\diamond[\![\alpha]\!] \leq f_\diamond[\![\alpha]\!]$ *and* $[\![\alpha]\!] \leq f_\diamond[\![\alpha]\!]$
(iv) $f_\blacklozenge f_\blacklozenge[\![\alpha]\!] \leq f_\blacklozenge[\![\alpha]\!]$ *and* $[\![\alpha]\!] \leq f_\blacklozenge[\![\alpha]\!]$
(v) $f_\diamond[\![\alpha]\!] \leq [\![\beta]\!]$ *iff* $[\![\alpha]\!] \leq f_\blacksquare[\![\beta]\!]$.
(vi) $f_\blacklozenge[\![\alpha]\!] \leq [\![\beta]\!]$ *iff* $[\![\alpha]\!] \leq f_\square[\![\beta]\!]$.
(vii) $f_\diamond[\![\alpha]\!] \wedge [\![\beta]\!] \leq f_\diamond([\![\alpha]\!] \wedge f_\blacksquare[\![\beta]\!])$ *and* $f_\blacklozenge[\![\alpha]\!] \wedge [\![\beta]\!] \leq f_\blacklozenge([\![\alpha]\!] \wedge f_\diamond[\![\beta]\!])$.

Proof. We only provide proof of (i). Let $f_\diamond[\![\alpha]\!] = [\![\gamma_1]\!]$, $f_\diamond[\![\beta]\!] = [\![\gamma_2]\!]$ and $f_\diamond[\![\alpha \vee \beta]\!] = [\![\gamma_3]\!]$ where $\gamma_1, \gamma_2, \gamma_3 \in \overline{T}$. Let $\vdash \gamma_3 \Rightarrow \theta$. Then $\vdash \diamond(\alpha \vee \beta) \Rightarrow \theta$. By Lemma 10, $\vdash \diamond\alpha \Rightarrow \theta$ and $\vdash \diamond\beta \Rightarrow \theta$. Hence $\vdash \gamma_1 \Rightarrow \theta$ and $\vdash \gamma_2 \Rightarrow \theta$. Thus by $(\vee L)$, one has $\vdash \gamma_1 \vee \gamma_2 \Rightarrow \theta$. Hence by Lemma 19, $[\![\gamma_3]\!] \leq [\![\gamma_1 \vee \gamma_2]\!]$, whence $f_\diamond[\![\alpha \vee \beta]\!] \leq f_\diamond[\![\alpha]\!] \vee f_\diamond[\![\beta]\!]$. Similarly $f_\diamond[\![\alpha]\!] \vee f_\diamond[\![\beta]\!] \leq f_\diamond[\![\alpha \vee \beta]\!]$. The proof for the second equality is analogous and hence is omitted. □

Theorem 21 $\mathcal{Q}$ *is a finite IS4t.*

Proof. By Lemma 15 and Lemma 20. □

Lemma 22 *Let $\alpha \wedge \beta$, $\alpha \to \beta$, $\diamond\alpha$, $\blacklozenge\alpha$, $\square\alpha$, $\blacksquare\alpha$, $\bot$, $\top \in T$, then the following equations hold:*

(i) $[\![\alpha \wedge \beta]\!] = [\![\alpha]\!] f_\wedge [\![\beta]\!]$.
(ii) $[\![\alpha \to \beta]\!] = [\![\alpha]\!] f_\to [\![\beta]\!]$.
(iii) $[\![\bot]\!] = f_0$ *and* $[\![\top]\!] = f_1$.
(iv) $[\![\diamond\alpha]\!] = f_\diamond[\![\alpha]\!]$ *and* $[\![\blacklozenge\alpha]\!] = f_\blacklozenge[\![\alpha]\!]$.
(v) $[\![\square\alpha]\!] = f_\square[\![\alpha]\!]$ *and* $[\![\blacksquare\alpha]\!] = f_\blacksquare[\![\alpha]\!]$.

Proof. We only provide proofs for (v). Clearly $\vdash \diamond\blacksquare\alpha \Rightarrow \alpha$. Hence, $f_\diamond[\![\blacksquare\alpha]\!] \leq [\![\alpha]\!]$. Thus $[\![\blacksquare\alpha]\!] \leq f_\blacksquare[\![\alpha]\!]$ by Lemma 20 (iii). Conversely let $\gamma_1, \ldots, \gamma_n \in T^{\#}$ and $f_\blacksquare[\![\alpha]\!] = [\![\gamma_1 \vee \ldots \vee \gamma_n]\!]$ where $f_\diamond[\![\gamma_i]\!] \leq \alpha$. Then $\vdash \diamond\gamma_i \Rightarrow \alpha$ for all $1 \leq i \leq n$. Hence by $(\blacksquare R)$ and $(\vee L)$, one gets $\vdash \gamma_1 \vee \ldots \vee \gamma_n \Rightarrow \blacksquare\alpha$. Therefore, $f_\blacksquare[\![\alpha]\!] \leq [\![\blacksquare\alpha]\!]$ since $\blacksquare\alpha \in T \subseteq \overline{T}$. □

Lemma 23 (FMP) *For any $\mathcal{RS}$-sequent $\alpha \Rightarrow \beta$, if $\nvdash \alpha \Rightarrow \beta$, then there exits a finite IS4t-algebra $\mathcal{Q}$, such that $\not\models_\mathcal{Q} \alpha \Rightarrow \beta$.*

Proof. Let T be the set of all subformulas of α and β. Let $T^{\#}$ be the $\wedge, \to$-closure of T. Obviously, $T^{\#}$ is closed under subformulas. Define $\overline{T}$ as above. Let $\sigma : \overline{T} \to |\overline{T}|$ be defined as $\sigma(p) = [\![p]\!]$ for all $p \in \mathbf{Var} \cap T$. By induction on the complexity of formula, Lemma 22 implies that $\sigma(\gamma) = [\![\gamma]\!]$ for any $\gamma \in T$.

Let $\mathcal{Q}$ be constructed as above and hence $\mathcal{Q}$ is finite. Assume $\models_{\mathcal{Q}} \alpha \Rightarrow \beta$ which implies that $\models_{\mathcal{Q}} \sigma(\alpha) \leq \sigma(\beta)$. By Theorem 9, $\vdash \alpha \Rightarrow \beta$. $\qquad \square$

Theorem 24 *Any sequents of $\mathcal{RS}$ of IS4t is decidable.*

4 Concluding Remarks

This paper contributes a decidable result of some particular form of formulas in intuitionistic tense logic S4. Certainly, this result might be extended to various variants of intuitionistic tense or modal logic, especially those whose decidability problems are still open.

References

[1] Aoto, T. and H. Shirasu, *On the finite model property of intuitionistic modal logics over MIPC*, Mathematical Logic Quarterly **45** (1999), p. 435–448.

[2] Balbiani, P. and M. Diéguez, *Temporal here and there*, in: *Logics in Artificial Intelligence*, Springer, 2016 pp. 81–96.

[3] Bull, R., *MIPC as the formalisation of an intuitionist concept of modality*, The Journal of Symbolic Logic **31** (1997), pp. 609–616.

[4] Chandra, A. K. and D. Harel, *Horn clause queries and generalizations*, The Journal of Logic Programming **2** (1985), pp. 1–15.

[5] Chen, C.-C. and I.-P. Lin, *The computational complexity of the satisfiability of modal Horn clauses for modal propositional logics*, Theoretical Computer Science **129** (1994), pp. 95–121.

[6] Davies, R., *A temporal logic approach to binding-time analysis*, Journal of the ACM (JACM) **64** (2017), pp. 1–45.

[7] de Paiva, V. and H. Eades III, *Constructive temporal logic, categorically*, IfCoLog Journal of Logics and their Applications (2017), p. 1287.

[8] Dzik, W. I., J. Järvinen and M. Kondo, *Intuitionistic propositional logic with galois connections*, Log. J. IGPL **18** (2010), pp. 837–858.

[9] Ewald, W. B., *Intuitionistic tense and modal logic*, The Journal of Symbolic Logic **51** (1986), pp. 166–179.

[10] Figallo, A. V. and G. Pelaitay, *An algebraic axiomatization of the Ewald's intuitionistic tense logic*, Soft Computing **18** (2014), pp. 1873–1883.

[11] Gelfond, M. and V. Lifschitz, *The stable model semantics for logic programming*, ICLP/SLP **88**, Cambridge, MA, 1988, pp. 1070–1080.

[12] Gelfond, M. and V. Lifschitz, *Classical negation in logic programs and disjunctive databases*, New Generation Computing **9** (1991), pp. 365–385.

[13] Girlando, M., R. Kuznets, S. Marin, M. Morales and L. Straßburger, *Intuitionistic s4 is decidable*, in: *2023 38th Annual ACM/IEEE Symposium on Logic in Computer Science (LICS)*, IEEE, 2023, pp. 1–13.

[14] Harrop, R., *On the existence of finite models and decision procedures for propositional calculi*, Mathematical Proceedings of the Cambridge Philosophical Society **54 (1)**, Cambridge University Press, 1958, pp. 1–13.

[15] Leone, N., G. Pfeifer, W. Faber, T. Eiter, G. Gottlob, S. Perri and F. Scarcello, *The DLV system for knowledge representation and reasoning*, ACM Transactions on Computational Logic (TOCL) **7** (2006), pp. 499–562.

[16] Liang, F. and Z. Lin, *On the decidability of intuitionistic tense logic without disjunction*, in: *Proceedings of the Twenty-Ninth International Conference on International Joint Conferences on Artificial Intelligence*, 2021, pp. 1798–1804.

[17] Lin, K. and Z. Lin, *The sequent systems and algebraic semantics of intuitionistic tense logics*, in: *International Workshop on Logic, Rationality and Interaction*, Lecture Notes in Computer Science **11813**, 2019, pp. 140–152.

[18] Maier, P., *Intuitionistic ltl and a new characterization of safety and liveness*, in: *International Workshop on Computer Science Logic*, Springer, 2004, pp. 295–309.

[19] Marek, V. W. and M. Truszczyński, *Stable models and an alternative logic programming paradigm*, in: *The Logic Programming Paradigm*, Springer, 1999 pp. 375–398.

[20] Nemitz, W. and T. Whaley, *Varieties of implicative semi-lattices. II*, Pacific Journal of Mathematics **45** (1973), pp. 303–311.

[21] Niemelä, I., *Logic programs with stable model semantics as a constraint programming paradigm*, Annals of Mathematics and Artificial Intelligence **25** (1999), pp. 241–273.

[22] Ono, H., *On some intuitionistic modal logics*, Publications of the Research Institute for Mathematical Sciences **13** (1977), pp. 687–722.

[23] Pearce, D., *A new logical characterisation of stable models and answer sets*, in: *International Workshop on Non-monotonic Extensions of Logic Programming*, 1996, pp. 57–70.

[24] Peng, Y., Z. Lin and F. Liang, *On the finite model property of weak intuitionistic tense logic*, in: *Logic, Rationality, and Interaction: 8th International Workshop*, Lecture Notes in Computer Science **13039**, 2021, pp. 174–182.

[25] Servi, G. F., *The finite model property for MIPQ and some consequences*, Notre Dame J. Formal Log. **19** (1978), pp. 687–692.

[26] Servi, G. F., *Axiomatizations for some intuitionistic modal logics*, Rendiconti Del Seminario Matematico **42** (1984), pp. 179–184.

[27] Simpson, A. K., "The proof theory and semantics of intuitionistic modal logic," Ph.D. thesis, University of Edinburgh, UK (1994).

[28] Wang, Y., Y. Peng and Z. Lin, *On the cut elimination of weak intuitionistic tense logic*, arXiv preprint arXiv:2405.09970 (2024).

[29] Wolter, F. and M. Zakharyaschev, *Intuitionistic modal logic*, in: A. Cantini, editor, *Logic and Foundations of Mathematics*, Kluwer Academic Publishers, 1999 pp. 227–238.

Model Theoretic Aspects of Modal Logic with Counting

Xiaoxuan Fu [1]

China University of Political Science and Law, Beijing, China

Zhiguang Zhao [2] [3]

Taishan University, Tai'an, China

Abstract

In this paper, we investigate the model theoretic aspects of modal logic with counting ML(#) and its variants. First of all, we show that for any natural number n, there is an ML(#)-formula φ that is satisfied only on models of size at least $\aleph_n$. Additionally, we prove that ML(#) with infinitely many propositional variables is not compact, nor is ML(#) with no variable and arbitrary depth. Furthermore, we present an example demonstrating the failure of interpolation for graded modal logic with counting GML(#) with respect to image-finite frames and finite frames. Finally, we give a proof of the Halldén completeness for the shallow fragment of ML(#) with respect to serial image-finite frames.

Keywords: modal logic with counting, compactness, interpolation, Halldén completeness

1 Introduction

The reasoning exhibited by individuals in their daily lives can be both qualitative and quantitative. Typically, logic pertains more to the qualitative aspect, while arithmetic/counting are more aligned with the quantitative aspect. Given that human reasoning often involves an interplay between qualitative and quantitative information, it becomes compelling to investigate the integration of logics with counting mechanisms. Specifically, comparisons between cardinalities represent interesting numerical information that frequently appears in daily reasoning.

---

[1] xfuuva@gmail.com

[2] zhaozhiguang23@gmail.com

[3] The research of the first author is supported by Tsinghua University Initiative Scientific Research Program. The research of the second author is supported by the Taishan Young Scholars Program of the Government of Shandong Province, China (No.tsqn201909151) and Shandong Provincial Natural Science Foundation, China (project number: ZR2023QF021).

In the existing literature, various works have explored the combination of cardinality comparisons with first-order logic [1,5,7,8]. However, the resulting logics are often of very high complexity and are typically not axiomatizable. This complexity makes it valuable to examine smaller languages, such as propositional logic with cardinality comparison formulas [3], monadic first-order logic with counting, modal logic with counting (ML(#)) [10], and modal logics with Presburger constraints [2]. These simpler logics, with their lower complexity, are particularly promising for applications in automated reasoning.

For AI, incorporating both qualitative and quantitative reasoning enhances the system's ability to emulate human-like decision-making processes. By the use of simpler logics with integrated counting mechanisms, AI can achieve more sophisticated and efficient reasoning capabilities. This advancement can significantly improve AI's performance in various complex tasks, making it a valuable area of research in the field of AI reasoning.

In this paper, we investigate some model-theoretic properties of ML(#) and its variants. First of all, we show that for any natural number n, there is an ML(#)-formula φ that is satisfied only on models of size at least $\aleph_n$. Additionally, we show that ML(#) is not compact by using the well-foundedness of sets of cardinal numbers. We also prove that GML(#) with respect to image-finite frames and finite frames do not have interpolation. Finally, we provide a proof that the shallow fragment of ML(#) on the class of serial image-finite frames has Halldén completeness.

The structure of the paper is as follows: Section 2 gives the preliminaries of ML(#). Section 3 shows that for any natural number n, there is an ML(#)-formula φ that is satisfied only on models of size at least $\aleph_n$. Section 4 proves that ML(#) is not compact by using the well-foundedness of sets of cardinal numbers. Section 5 proves that GML(#) with respect to image-finite frames and finite frames do not have interpolation. In Section 6, we give the proof of that the shallow fragment of ML(#) on the class of serial image-finite frames has Halldén completeness. Section 7 gives conclusions.

2 Preliminaries

In the present section, we give preliminaries on the ML(#). For more details, see Section 7 in [10].

Syntax Given a countable set Prop of propositional variables, we define the formulas and numerical terms of ML(#) as follows:

$$\text{formulas:} \quad p \mid \bot \mid \top \mid \neg\varphi \mid \varphi \wedge \psi \mid \#\varphi \succsim \#\psi$$
$$\text{numerical terms:} \quad \#\varphi$$

where $p \in$ Prop. We use standard abbreviations for $\vee, \rightarrow, \leftrightarrow$. We use $\boldsymbol{p}$ to denote a tuple of propositional variables like $(p_1, \ldots, p_n)$.

Definition 2.1 [Counting depth] The counting depth of formulas is defined recursively as follows:

- $d(p) = d(\bot) = d(\top) = 0$;

- $d(\neg\varphi) = d(\varphi)$;
- $d(\varphi \wedge \psi) = max\{d(\varphi), d(\psi)\}$;
- $d(\#\varphi \succsim \#\psi) = max\{d(\varphi), d(\psi)\} + 1$.

In Section 6, we will make use of the following definition of shallow formulas, which will be used to define subsets of natural numbers:

Definition 2.2 [Shallow formulas] We say that an ML(#)-formula is shallow if it is a Boolean combination of formulas of the form $\#\varphi \succsim \#\psi$ where φ, ψ are of counting depth 0.

Semantics ML(#)-formulas are interpreted on Kripke frames $\mathbb{F} = (W, R)$ where $W \neq \emptyset$ is the domain and R is a binary relation on W. A Kripke model is a tuple $\mathbb{M} = (\mathbb{F}, V)$ where $V : \mathsf{Prop} \to \mathsf{P}(W)$ is a valuation on W. We use $R_s = \{t : Rst\}$ to denote the set of successors of s. A Kripke frame is:

- finite, if W is finite;
- image-finite, if R_s is finite for every $s \in W$;
- serial, if R_s is non-empty for every $s \in W$.

We use $[\![\varphi]\!]^{\mathbb{M}}$ to denote the set of worlds in $\mathbb{M}$ where φ is true. The satisfaction relation for the basic case and Boolean connectives are defined as usual. For numerical terms,

$$[\![\#\varphi]\!]^{\mathbb{M},s} = |R_s \cap [\![\varphi]\!]^{\mathbb{M}}|,$$

i.e. $[\![\#\varphi]\!]^{\mathbb{M},s}$ is the number of successors of s where φ is true.
For cardinality comparison formulas,

$$\mathbb{M}, s \Vdash \#\varphi \succsim \#\psi \text{ iff } [\![\#\varphi]\!]^{\mathbb{M},s} \geq [\![\#\psi]\!]^{\mathbb{M},s}$$

i.e. $\#\varphi \succsim \#\psi$ is true at s if more (or the same number of) R-successors of s make φ true than making ψ true.

Some abbreviations We define the following abbreviations:

- $\#\varphi \succ \#\psi$ is defined as $(\#\varphi \succsim \#\psi) \wedge \neg(\#\psi \succsim \#\varphi)$;
- $\#\varphi = \#\psi$ is defined as $(\#\varphi \succsim \#\psi) \wedge (\#\psi \succsim \#\varphi)$;
- The standard modality $\Diamond\varphi$ is defined as $\#\varphi \succ \#\bot$;
- $\Box\varphi$ is defined as $\neg\Diamond\neg\varphi$.

Proposition 2.3 *The following equivalences hold on the basis of the semantics of $\#\varphi \succsim \#\psi$:*

- $\mathbb{M}, s \Vdash \#\varphi \succ \#\psi$ *iff* $[\![\#\varphi]\!]^{\mathbb{M},s} > [\![\#\psi]\!]^{\mathbb{M},s}$;
- $\mathbb{M}, s \Vdash \#\varphi = \#\psi$ *iff* $[\![\#\varphi]\!]^{\mathbb{M},s} = [\![\#\psi]\!]^{\mathbb{M},s}$;
- $\mathbb{M}, s \Vdash \Diamond\varphi$ *iff* $[\![\#\varphi]\!]^{\mathbb{M},s} > 0$ *iff there exists* $t \in W$ *such that* Rst *and* $\mathbb{M}, t \Vdash \varphi$;
- $\mathbb{M}, s \Vdash \Box\varphi$ *iff for any* $t \in W$ *such that* Rst, *we have* $\mathbb{M}, t \Vdash \varphi$.

Propositional quantifiers In the present paper, we will use existential propositional quantifiers of the form $\exists p$ (where p is a propositional variable) to talk about the existence of a valuation such that a certain ML(#)-formula is satisfiable under this valuation.

We use V_X^p to denote a valuation which is the same as V except that $V_X^p(p) = X \subseteq W$. The additional satisfaction relation clauses are defined as follows:

$$\mathbb{M}, w \Vdash \forall p\varphi \quad \text{iff} \quad \text{for all } X \subseteq W, (W, R, V_X^p), w \Vdash \varphi;$$
$$\mathbb{M}, w \Vdash \exists p\varphi \quad \text{iff} \quad \text{there exists } X \subseteq W \text{ such that } (W, R, V_X^p), w \Vdash \varphi.$$

Graded modal logic with counting GML(#) In graded modal logic [11,6], we have graded modalities $\Diamond_{\geq n}\varphi$ for each positive natural number n, intuitively reads "there are at least n successors satisfying φ". It is natural to consider the graded extension of modal logic with counting GML(#). The following abbreviations are used:

- $\Diamond_{\leq n}\varphi$ is defined as $\neg\Diamond_{\geq n+1}\varphi$;

- $\Diamond_{=n}\varphi$ is defined as $\Diamond_{\geq n}\varphi \wedge \Diamond_{\leq n}\varphi$.

For counting depth, we define $d(\Diamond_{\geq n}\varphi) = d(\varphi) + 1$.

For the semantics of $\Diamond_{\geq n}\varphi$,

$$\mathbb{M}, s \Vdash \Diamond_{\geq n}\varphi \text{ iff } [\![\#\varphi]\!]^{\mathbb{M},s} \geq n.$$

The Power of definability of ML(#) We briefly recall the results in [4] which will be used in Section 6 on Halldén completeness. Here $\mathbb{N}$ denotes the set of natural numbers.

Definition 2.4 Given a subset $X \subseteq \mathbb{N}$, a shallow ML(#)-formula φ with all propositional variables occurring in $\boldsymbol{p}$ defines X if for all image-finite pointed Kripke frames $(\mathbb{F}, s)$,

$$\mathbb{F}, s \Vdash \exists \boldsymbol{p}\varphi \text{ iff } |R_s| \in X.$$

The characterization of definable subsets will make use of the following two definitions:

Definition 2.5 [Semilinear sets] A subset $X \subseteq \mathbb{N}$ is said to be linear if it is of the form $X = \{a + b_1 \cdot x_1 + \ldots + b_n \cdot x_n \mid x_1, \ldots, x_n \in \mathbb{N}\}$ for some fixed $a, b_1, \ldots, b_n \in \mathbb{N}$. A subset $X \subseteq \mathbb{N}$ is said to be semilinear if it is a finite union of linear subsets.

Definition 2.6 We say that a subset $X \subseteq \mathbb{N}$ is closed under taking multiples, if for any $n \in X$ and $2 \leq m \in \mathbb{N}$, we have that $m \cdot n \in X$.

The characterization theorem of definable subsets of natural numbers is given as follows:

Theorem 2.7 (Section 4 in [4]) *For any subset $X \subseteq \mathbb{N}$, X is definable by a shallow ML(#)-formula iff it is semilinear and closed under taking multiples.*

3 Model size

In this section, we show that for any natural number n, there is an ML(#)-formula φ that is satisfied only on models of size at least $\aleph_n$.

3.1 Formula that only has models of size at least $\aleph_0$

Proposition 3.1 (Proposition 6 in [4]) $\mathbb{F}, s \Vdash \exists p((\#p = \#\top) \wedge \Diamond \neg p)$ *iff* $|R_s| \geq \aleph_0$.

Proof. ($\Leftarrow$): Suppose that in the pointed frame $(\mathbb{F}, s)$, s has infinitely many successors. Then take $t \in R_s$, we can define a valuation V on $\mathbb{F}$ such that $V(p) = R_s - \{t\}$. Then $|R_s - \{t\}| = |R_s|$, so $|R_s \cap [\![p]\!]^{\mathbb{F},V,s}| = |R_s \cap [\![\top]\!]^{\mathbb{F},V,s}|$, so $\mathbb{F}, V, s \Vdash \#p = \#\top$. Since $\mathbb{F}, V, t \Vdash \neg p$, we have that $\mathbb{F}, V, s \Vdash \Diamond \neg p$. So $\mathbb{F}, s \Vdash \exists p((\#p = \#\top) \wedge \Diamond \neg p)$.

($\Rightarrow$): Suppose $\mathbb{F}, s \Vdash \exists p((\#p = \#\top) \wedge \Diamond \neg p)$. Then there is a valuation V on $\mathbb{F}$ such that $\mathbb{F}, V, s \Vdash (\#p = \#\top) \wedge \Diamond \neg p$. So there is a successor t of s such that $\mathbb{F}, V, t \Vdash \neg p$ and $|R_s \cap [\![p]\!]^{\mathbb{F},V,s}| = |R_s \cap [\![\top]\!]^{\mathbb{F},V,s}| = |R_s|$. So $R_s \cap [\![p]\!]^{\mathbb{F},V,s}$ is a proper subset of R_s with the same cardinality. So s has infinitely many successors. $\square$

From this proof, we can see that if $(\#p = \#\top) \wedge \Diamond \neg p$ is satisfiable at a model $(\mathbb{M}, s) = (W, R, V, s)$, then $W, R, s \Vdash \exists p((\#p = \#\top) \wedge \Diamond \neg p)$, so $|R_s| \geq \aleph_0$, which forces the frame to have size at least $\aleph_0$.

3.2 Formula that only has models of size at least $\aleph_1$

By an argument similar to the previous part, we have the following proposition:

Proposition 3.2 (Proposition 7 in [4]) *Suppose that p does not occur in φ. Then* $\mathbb{M}, s \Vdash \exists p((\#(p \wedge \varphi) = \#\varphi) \wedge \Diamond(\neg p \wedge \varphi))$ *iff* $[\![\#\varphi]\!]^{\mathbb{M},s} \geq \aleph_0$.

We denote $\exists p((\#(p \wedge \varphi) = \#\varphi) \wedge \Diamond(\neg p \wedge \varphi))$ as $\exists^{\geq \aleph_0}\varphi$.

Then we have the following result, which forces the number of successors to be at least $\aleph_1$:

Proposition 3.3 (Proposition 8 in [4]) $\mathbb{F}, s \Vdash \exists q \exists r(\exists^{\geq \aleph_0}(q \wedge r) \wedge (\#(q \wedge \neg r) \succ \#(q \wedge r)))$ *iff* $|R_s| \geq \aleph_1$.

Proof. ($\Leftarrow$): Suppose that in the pointed frame $(\mathbb{F}, s)$, s has at least $\aleph_1$ successors. Take disjoint $X, Y \subseteq R_s$ such that $|X| = \aleph_0$, $|Y| = \aleph_1$, then define a valuation V on $\mathbb{F}$ such that all points in X make $q \wedge r$ true and all points in Y make $q \wedge \neg r$ true. Then $\mathbb{F}, V, s \Vdash \exists^{\geq \aleph_0}(q \wedge r)$ and $\mathbb{F}, V, s \Vdash \#(q \wedge \neg r) \succ \#(q \wedge r)$, So $\mathbb{F}, s \Vdash \exists q \exists r(\exists^{\geq \aleph_0}(q \wedge r) \wedge (\#(q \wedge \neg r) \succ \#(q \wedge r)))$.

($\Rightarrow$): Suppose $\mathbb{F}, s \Vdash \exists q \exists r(\exists^{\geq \aleph_0}(q \wedge r) \wedge (\#(q \wedge \neg r) \succ \#(q \wedge r)))$. Then there is a valuation V on $\mathbb{F}$ such that $\mathbb{F}, V, s \Vdash \exists^{\geq \aleph_0}(q \wedge r)$ and $\mathbb{F}, V, s \Vdash \#(q \wedge \neg r) \succ \#(q \wedge r)$. Therefore, there are at least $\aleph_0$ many successors of s satisfying $q \wedge r$ and the cardinality of s' successors with $q \wedge \neg r$ true is strictly larger than the cardinality of those with $q \wedge r$ true. Therefore there are at least $\aleph_1$ many successors of s with $q \wedge \neg r$ true. So $|R_s| \geq \aleph_1$. $\square$

Since $\exists q \exists r(\exists^{\geq \aleph_0}(q \wedge r) \wedge (\#(q \wedge \neg r) \succ \#(q \wedge r)))$ is equivalent to $\exists p \exists q \exists r(((\#(p \wedge q \wedge r) = \#(q \wedge r)) \wedge \Diamond(\neg p \wedge q \wedge r)) \wedge (\#(q \wedge \neg r) \succ \#(q \wedge r)))$, so consider the ML($\#$)-formula

$$((\#(p \wedge q \wedge r) = \#(q \wedge r)) \wedge \Diamond(\neg p \wedge q \wedge r)) \wedge (\#(q \wedge \neg r) \succ \#(q \wedge r)),$$

if it is satisfiable at a model $(\mathbb{M}, s) = (W, R, V, s)$, then $|R_s| \geq \aleph_1$, which forces the frame to have size at least $\aleph_1$.

3.3 Formulas that only have models of size at least $\aleph_n$

Now we are ready to define the formulas which only have models of size at least $\aleph_n$, for any $n \in \mathbb{N}$:

For any $n \in \mathbb{N}^+$, consider the smallest natural number k such that $2^k \geq n + 1$. For propositional variables $p_1, \ldots, p_k$, list all the conjunctive clauses of the form $(\neg)p_1 \wedge \ldots \wedge (\neg)p_k$ as $S_1, \ldots, S_{2^k}$. It is easy to see that these conjunctive clauses are pairwise inconsistent, and their disjunction is a tautology.

Now by an argument similar to Proposition 3.3, we have the following result:

Proposition 3.4 $\mathbb{F}, s \Vdash \exists p_1 \ldots \exists p_k (\exists^{\geq \aleph_0} S_1 \wedge (\#S_{n+1} \succ \#S_n \succ \ldots \succ \#S_1))$ iff $|R_s| \geq \aleph_n$.

Now since $\exists p_1 \ldots \exists p_k (\exists^{\geq \aleph_0} S_1 \wedge (\#S_{n+1} \succ \#S_n \succ \ldots \succ \#S_1))$ is equivalent to $\exists p \exists p_1 \ldots \exists p_k ((\#(p \wedge S_1) = \#S_1) \wedge \Diamond(\neg p \wedge S_1) \wedge (\#S_{n+1} \succ \#S_n \succ \ldots \succ \#S_1))$, consider the ML(#)-formula

$$(\#(p \wedge S_1) = \#S_1) \wedge \Diamond(\neg p \wedge S_1) \wedge (\#S_{n+1} \succ \#S_n \succ \ldots \succ \#S_1),$$

if it is satisfiable at a model $(\mathbb{M}, s) = (W, R, V, s)$, then $|R_s| \geq \aleph_n$, which forces the frame to have size at least $\aleph_n$.

4 Failure of compactness

In Proposition 1 in [10], van Benthem and Icard show that first-order logic with counting FO(#) lacks compactness, by an argument using the ability to define $\exists^\infty x.\varphi$ stating that there are infinitely many x's satisfying φ and consider the set $\{\exists^{\geq n} x.Px \mid n \in \mathbb{N}\} \cup \{\neg \exists^\infty x.Px\}$.

In the present section, we show that ML(#) is not compact when we have infinitely many propositional variables, and even if we have no propositional variable, if we allow arbitrary depth of formulas, then ML(#) is also not compact. The basic proof strategy is to use the property of cardinal numbers that sets of cardinal numbers are well-founded.

Theorem 4.1 *Consider ML(#) with infinitely many propositional variables. Then there are formulas $\{\varphi_i\}_{i \in \mathbb{N}}$ such that they are finitely satisfiable but not satisfiable together. Indeed, we can take all φ_i's to be shallow formulas.*

Proof. Consider the following formulas:
$\varphi_1 := \#p_1 \succ \#p_2$;
$\varphi_2 := \#p_1 \succ \#p_2 \succ \#p_3$;
$\varphi_3 := \#p_1 \succ \#p_2 \succ \#p_3 \succ \#p_4$;
$\vdots$

Then for any finitely many formulas $\varphi_{i_1}, \ldots, \varphi_{i_k}$, suppose the largest index is n, then their conjunction is equivalent to $\#p_1 \succ \#p_2 \succ \ldots \succ \#p_{n+1}$. Then consider the pointed Kripke model $(\mathbb{M}, w_0) = (W, R, V, w_0)$ where $W = \{w_0, \ldots, w_{n+1}\}$, $R = \{(w_0, w_i) \mid 1 \leq i \leq n+1\}$, $V(p_i) = \{w_j \mid i \leq j \leq n+1\}$.

Then $[\![\#p_i]\!]^{\mathbb{M},w_0} = n+2-i$ and $[\![\#p_1]\!]^{\mathbb{M},w_0} > [\![\#p_2]\!]^{\mathbb{M},w_0} > \ldots > [\![\#p_{n+1}]\!]^{\mathbb{M},w_0}$, so $\varphi_{i_1} \wedge \ldots \wedge \varphi_{i_k}$ is satisfiable.

However, if we consider $\{\varphi_i\}_{i \in \mathbb{N}}$, if this set is satisfiable, then there is a pointed Kripke model $(\mathbb{M}, w_0)$ such that $[\![\#p_1]\!]^{\mathbb{M},w_0} > [\![\#p_2]\!]^{\mathbb{M},w_0} > \ldots$, which means that there is an infinite descending chain of cardinal numbers, a contradiction.

Notice that in the formulas above, we only use shallow formulas, although we use infinitely many propositional variables. $\qquad\square$

Indeed, if we use formulas of arbitrary finite depth, then even if we have no propositional variable, the resulting fragment is already not compact, which is shown in the next theorem.

Theorem 4.2 *Consider $ML(\#)$ with no propositional variable and formulas of arbitary depth are allowed. Then there are formulas $\{\varphi_i\}_{i \in \mathbb{N}}$ such that they are finitely satisfiable but not satisfiable together.*

Proof. Consider the following formulas:

$\varphi_1 := \#\top \succ \#\Diamond\top$;
$\varphi_2 := \#\top \succ \#\Diamond\top \succ \#\Diamond\Diamond\top$;
$\varphi_3 := \#\top \succ \#\Diamond\top \succ \#\Diamond\Diamond\top \succ \#\Diamond\Diamond\Diamond\top$;
$\vdots$

Then for any finitely many formulas $\varphi_{i_1}, \ldots, \varphi_{i_k}$, suppose the largest index is n, then their conjunction is equivalent to $\#\top \succ \#\Diamond\top \succ \ldots \succ \#\Diamond^n\top$. We construct the following pointed Kripke model $(\mathbb{M}, w_0)$:

- $W = \{w_0\} \cup \{w_{i,j} \mid 1 \le j \le i \le n\}$;
- $R = \{(w_0, w_{i,1}) \mid 1 \le i \le n\} \cup \{(w_{i,j}, w_{i,k}) \mid 1 \le j < k \le i \le n \text{ and } k = j+1\}$;
- V is the vacuous valuation (since there is no propositional variable).

Then $[\![\#\Diamond^i\top]\!]^{\mathbb{M},w_0} = n - i$ where $0 \le i \le n$, so $[\![\#\top]\!]^{\mathbb{M},w_0} > [\![\#\Diamond\top]\!]^{\mathbb{M},w_0} > \ldots > [\![\#\Diamond^n\top]\!]^{\mathbb{M},w_0}$, therefore $\varphi_{i_1} \wedge \ldots \wedge \varphi_{i_k}$ is satisfiable.

However, if we consider $\{\varphi_i\}_{i \in \mathbb{N}}$, if this set is satisfiable, then there is a pointed Kripke model $(\mathbb{M}, w)$ such that $[\![\#\top]\!]^{\mathbb{M},w_0} > [\![\#\Diamond\top]\!]^{\mathbb{M},w_0} > [\![\#\Diamond\Diamond\top]\!]^{\mathbb{M},w_0} > \ldots$, which means that there is an infinite descending chain of cardinal numbers, a contradiction. $\qquad\square$

Remark 4.3 From the construction of the counterexamples, we can see that if a cardinality comparison language involves infinitely many non-equivalent formulas, then we can construct the infinite descending chain of cardinal numbers, so the language cannot be compact. However, when we only have finitely many variables and finite depth, then $ML(\#)$ has to be compact since there are only finitely many non-equivalent formulas, according to the normal form theorem in Section 7.4 in [10].

5 Interpolation failure for GML^{fin}(#)

In this section, we use an argument similar to Proposition 2 in Section 3.3.1 in [10] to show that graded modal logic with counting on the class of image-finite frames and the class of finite frames does not have interpolation. Since the proofs for the two cases are essentially the same, we only prove it for the case of image-finite frames.

Fist of all, we consider the following definability result which will be used in the proof of interpolation failure. Notice that the set of odd numbers is not closed under taking multiples (see Definition 2.6) so it cannot be define by a shallow ML(#)-formula.

Lemma 5.1 *In the class of image-finite frames,*

- *$\#p = \#\neg p$ is satisfiable at $(\mathbb{F}, s)$ iff s has an even number of successors.*
- *$\psi := \#(q \wedge r) = \#(q \wedge \neg r) \wedge \Diamond_{=1}\neg q$ is satisfiable at $(\mathbb{F}, s)$ iff s has an odd number of successors.*

Proof.

- For $\#p = \#\neg p$, the proof is similar to Proposition 2 in [4].

  ($\Leftarrow$): Suppose that s has an even number of successors. Then we can divide R_s into two disjoint parts X and Y of equal cardinality. Consider a valuation V such that $V(p) = X$, then $\mathbb{F}, V, s \Vdash \#p = \#\neg p$.

  ($\Rightarrow$): Suppose there is a valuation V on $\mathbb{F}$ such that $\mathbb{F}, V, s \Vdash \#p = \#\neg p$, i.e. $[\![\#p]\!]^{\mathbb{M},s} = [\![\#\neg p]\!]^{\mathbb{M},s}$, i.e. $|R_s \cap [\![p]\!]^{\mathbb{M}}| = |R_s \cap [\![\neg p]\!]^{\mathbb{M}}|$. Since $R_s = (R_s \cap [\![p]\!]^{\mathbb{M}}) \cup (R_s \cap [\![\neg p]\!]^{\mathbb{M}})$, we have that R_s can be divided into two disjoint parts of equal cardinality. Therefore, s has an even number of successors.

- ($\Leftarrow$): Suppose that s has an odd number of successors. Then we can divide R_s into three disjoint parts X, Y, Z, such that X and Y have the same cardinality and Z is a singleton. Consider a valuation V such that $V(q) = X \cup Y$, $V(r) = X$, then $V(q \wedge r) = X$, $V(q \wedge \neg r) = Y$, $V(\neg q) = Z$, so $\mathbb{F}, V, s \Vdash \#(q \wedge r) = \#(q \wedge \neg r) \wedge \Diamond_{=1}\neg q$.

  ($\Rightarrow$): Suppose there is a valuation V on $\mathbb{F}$ such that $\mathbb{F}, V, s \Vdash \#(q \wedge r) = \#(q \wedge \neg r) \wedge \Diamond_{=1}\neg q$. Then $[\![\#(q \wedge r)]\!]^{\mathbb{M},s} = [\![\#(q \wedge \neg r)]\!]^{\mathbb{M},s}$ and $[\![\#\neg q]\!]^{\mathbb{M},s} = 1$, i.e. $|R_s \cap [\![(q \wedge r)]\!]^{\mathbb{M}}| = |R_s \cap [\![(q \wedge \neg r)]\!]^{\mathbb{M}}|$ and $|R_s \cap [\![\neg q]\!]^{\mathbb{M}}| = 1$. Therefore, R_s can be divided into three disjoint parts X, Y, Z, such that X and Y have the same cardinality and Z is a singleton. Thus s has an odd number of successors.

$\square$

The following lemma is an indistinguishability result, which says that GML(#)-formulas without propositional variables and with grade number $\leq n$ cannot distinguish point with $n+5$ successors from point with $n+6$ successors:

Lemma 5.2 *Consider the following two frames: $\mathbb{F}$ with one root s and $n + 5$ dead end successors, and $\mathbb{G}$ with one root t and $n+6$ dead end successors. Then for any graded modal logic with counting GML(#)-formula θ built up from no propositional variable but only $\top$, $\bot$ using cardinality comparison formulas and*

111

graded modalities of grade number $\leq n$ (i.e., the largest number occurring in a graded modality $\Diamond_{\geq n}$), we have that $(\mathbb{F}, s) \Vdash \theta$ iff $(\mathbb{G}, t) \Vdash \theta$.

Proof. First of all, notice that we do not need a valuation V because we have no propositional variable.

For any formula φ in GML($\#$) without propositional variables, among the dead end successors, the formula is either true in all such successors or false in all such successors, so for any φ, either $[\![\#\varphi]\!]^{\mathbb{F},s} = n + 5$ and $[\![\#\varphi]\!]^{\mathbb{G},t} = n + 6$, or $[\![\#\varphi]\!]^{\mathbb{F},s} = [\![\#\varphi]\!]^{\mathbb{G},t} = 0$.

Now we can prove by induction on θ showing that $(\mathbb{F}, s) \Vdash \theta$ iff $(\mathbb{G}, t) \Vdash \theta$:

- When θ is $\bot$ or $\top$, trivial. The Boolean cases are also easy.

- When θ is $\Diamond_{\geq m}\gamma$, then since $1 \leq m \leq n$, we have

$$
\begin{array}{rl}
& (\mathbb{F}, s) \Vdash \theta \\
\text{iff} & [\![\#\gamma]\!]^{\mathbb{F},s} \geq m \\
\text{iff} & [\![\#\gamma]\!]^{\mathbb{F},s} = n + 5 \\
\text{iff} & [\![\#\gamma]\!]^{\mathbb{G},t} = n + 6 \\
\text{iff} & [\![\#\gamma]\!]^{\mathbb{G},t} \geq m \\
\text{iff} & (\mathbb{G}, t) \Vdash \theta.
\end{array}
$$

- When θ is $\#\gamma \succsim \#\delta$, then

$$
\begin{array}{rl}
& (\mathbb{F}, s) \nVdash \theta \\
\text{iff} & (\mathbb{F}, s) \Vdash \#\delta \succ \#\gamma \\
\text{iff} & [\![\#\delta]\!]^{\mathbb{F},s} = n + 5 \text{ and } [\![\#\gamma]\!]^{\mathbb{F},s} = 0 \\
\text{iff} & [\![\#\delta]\!]^{\mathbb{G},t} = n + 6 \text{ and } [\![\#\gamma]\!]^{\mathbb{G},t} = 0 \\
\text{iff} & (\mathbb{G}, t) \Vdash \#\delta \succ \#\gamma \\
\text{iff} & (\mathbb{G}, t) \nVdash \theta.
\end{array}
$$

$\square$

Now we have the following main result:

Theorem 5.3 *Graded modal logic with counting on the class of image-finite frames does not have interpolation.*

Proof. We consider the formulas $\varphi := \#p = \#\neg p$, $\psi := \#(q \wedge r) = \#(q \wedge \neg r) \wedge \Diamond_{=1}\neg q$. We will show that $\Vdash \varphi \to \neg\psi$ but φ and $\neg\psi$ have no interpolant.

To check that $\Vdash \varphi \to \neg\psi$, it suffices to check that $\Vdash \exists p\varphi \to \forall q\forall r\neg\psi$. According to the lemma before, $(\mathbb{F}, s) \Vdash \exists p\varphi$ iff s has an even number of successors, and $(\mathbb{F}, s) \Vdash \forall q\forall r\neg\psi$ iff s does not have an odd number of successors, in image-finite frames this means that s has an even number of successors. Therefore $\Vdash \varphi \to \neg\psi$ holds.

If an interpolant θ without any propositional variable exists for φ and $\neg\psi$, then $\Vdash \varphi \to \theta$ and $\Vdash \theta \to \neg\psi$, so $\Vdash \exists p\varphi \to \theta$ and $\Vdash \theta \to \forall q\forall r\neg\psi$, which means that θ is true at $(\mathbb{F}, s)$ iff s has an even number of successors.

Now we show that θ is not definable in graded modal logic with counting GML($\#$) in image-finite frames if we do not have any propositional variable but $\top$ and $\bot$ only:

Suppose there is such a θ built up from $\top$ and $\bot$ using connectives in

GML($\#$), then there is an upper bound in the grade number in the formula θ, say n (i.e., the largest number occurring in a graded modality $\Diamond_{\geq n}$). By Lemma 5.2, for $\mathbb{F}$ with one root s and $n+5$ dead end successors, and $\mathbb{G}$ with one root t and $n+6$ dead end successors, $(\mathbb{F}, s) \Vdash \theta$ iff $(\mathbb{G}, t) \Vdash \theta$. But θ can distinguish even number of successors and odd number of successors, so θ have different values in $(\mathbb{F}, s)$ and $(\mathbb{G}, t)$, a contradiction.

Therefore, such a θ cannot be defined in GML($\#$). So graded modal logic with counting on the class of image-finite frames does not have interpolation.□

Remark 5.4 For the result that we proved here, we notice that both the usage of graded modal logic with counting and the condition of image-finiteness play an important role. The usage of graded modalities guarantees that we can define a semilinear set which is not closed under taking multiples (i.e. the set of odd numbers), therefore we can use $\exists q \exists r \psi$ to define the set of odd numbers and $\forall q \forall r \neg \psi$ to define the set of even numbers. If we only have modal logic with counting ML($\#$), then we cannot define the set of even numbers using universally quantified formulas, since existentially quantified formulas cannot define the set of odd numbers (due to that the set of odd numbers is not closed under taking multiples).

For the requirement of image-finiteness, consider the formulas φ and ψ, they are both satisfiable at nodes with infinitely many successors, therefore if we drop the condition of image-finiteness, then $\exists p \varphi$ defines the class of even numbers and infinite cardinals, and $\exists q \exists r \psi$ defines the class of odd numbers and infinite cardinals. Apparently this is problematic, since in this case $\Vdash \exists p \varphi \rightarrow \forall q \forall r \neg \psi$ does not hold anymore.

6 Halldén completeness for the shallow fragment of $\mathbf{D}^{\text{fin}}(\#)$

In order to meet some certain constructivist demands, for logics like intuitionistic logics and relevance logics, their systems expect for the full disjunction property. Halldén completeness is a weaker property, in which there is growing interest within both philosophy and computer science. A logic L is Halldén-complete if, for any formulas φ and ψ having no variables in common, $\varphi \lor \psi \in L$ implies that $\varphi \in L$ or $\psi \in L$. In [9], van Benthem and Humberstone give a sufficient condition for normal modal logics to be Halldén-complete.

In this section, we prove that the shallow fragment (i.e., the fragment of all the shallow formulas) of modal logic with counting on the class of serial image-finite frames has Halldén completeness. In this proof, we make use of Theorem 2.7 which is proved in Section 4 in [4].

Theorem 6.1 *If two shallow formulas φ and ψ share no propositional variables and $\varphi \lor \psi$ is $D^{\text{fin}}(\#)$-valid (i.e. valid on the class of serial image-finite frames), then either φ is $D^{\text{fin}}(\#)$-valid or ψ is $D^{\text{fin}}(\#)$-valid.*

Proof. Suppose $\varphi \lor \psi$ is $\mathrm{D}^{\text{fin}}(\#)$-valid, then $\forall \boldsymbol{p} \varphi \lor \forall \boldsymbol{q} \psi$ is $\mathrm{D}^{\text{fin}}(\#)$-valid, where $\boldsymbol{p}$ are all the propositional variables occurring in φ and $\boldsymbol{q}$ are all the propositional variables occurring in ψ. Therefore, for any image-finite serial frame $\mathbb{F} =$

(W, R), any valuation V, any $w \in W$, $\mathbb{F}, V, w \Vdash \forall \boldsymbol{p}\varphi \vee \forall \boldsymbol{q}\psi$, so $\mathbb{F}, V, w \Vdash \forall \boldsymbol{p}\varphi$ or $\mathbb{F}, V, w \Vdash \forall \boldsymbol{q}\psi$, i.e. $\mathbb{F}, V, w \nVdash \exists \boldsymbol{p}\neg\varphi$ or $\mathbb{F}, V, w \nVdash \exists \boldsymbol{q}\neg\psi$. Since the valuation here plays no role, we have that $\mathbb{F}, w \nVdash \exists \boldsymbol{p}\neg\varphi$ or $\mathbb{F}, w \nVdash \exists \boldsymbol{q}\neg\psi$.

By Theorem 2.7, there are two subsets Y_1 and Y_2 of $\mathbb{N}$ such that Y_1 and Y_2 are defined by $\neg\varphi$ and $\neg\psi$, respectively.

Therefore, we have that for any serial image-finite frame $\mathbb{F} = (W, R)$, any $w \in W$, $|R_w| \notin Y_1$ or $|R_w| \notin Y_2$. Since $|R_w|$ ranges over all positive natural numbers, we have that $Y_1 \cap Y_2 \subseteq \{0\}$. Since Y_1 and Y_2 are both semilinear and closed under multiples, we have that if Y_1 and Y_2 both have non-zero elements k and l respectively, then $k \cdot l$ belong to both, a contradiction. So $Y_1 \subseteq \{0\}$ or $Y_2 \subseteq \{0\}$.

When $Y_1 \subseteq \{0\}$, we have that for any serial image-finite frame $\mathbb{F} = (W, R)$, any $w \in W$, $|R_w| \neq 0$, so $|R_w| \notin Y_1$, so $\mathbb{F}, w \nVdash \exists \boldsymbol{p}\neg\varphi$, i.e. $\mathbb{F}, w \Vdash \forall \boldsymbol{p}\varphi$. So φ is $\mathrm{D}^{\mathsf{fin}}(\#)$-valid. If $Y_2 \subseteq \{0\}$, then by a similar argument we have that ψ is $\mathrm{D}^{\mathsf{fin}}(\#)$-valid. So either φ is $\mathrm{D}^{\mathsf{fin}}(\#)$-valid or ψ is $\mathrm{D}^{\mathsf{fin}}(\#)$-valid. □

7 Conclusion

In this paper, we investigate some model-theoretic properties of $\mathrm{ML}(\#)$ and its variants. For further directions, we mention the followings:

- For the model size, it is easy to see that using cardinality comparison formulas, we can force the number of successors to have cardinality $\geq \aleph_n$ for any $n \in \mathbb{N}$, but if we allow countable conjunctions and disjunctions, the forcing power is not clear.

- For interpolation, it is not clear whether $\mathrm{ML}(\#)$ lacks interpolation, since we cannot apply the same kind of construction as in Proposition 2 in [10]. It is clear that the interpolant lives in second-order modal logic with counting/second-order graded modal logic with counting, but it is not clear if we can we find smaller language that have interpolation.

References

[1] Antonelli, G. A., *Numerical abstraction via the Frege quantifier*, Notre Dame Journal of Formal Logic **51** (2010), pp. 161–179.

[2] Demri, S. and D. Lugiez, *Complexity of modal logics with presburger constraints*, Journal of Applied Logic **8** (2010), pp. 233–252.

[3] Ding, Y., M. Harrison-Trainor and W. H. Holliday, *The logic of comparative cardinality*, Journal of Symbolic Logic **85** (2020), pp. 972–1005.

[4] Fu, X. and Z. Zhao, *Numerical expressive power of logical languages with cardinality comparison*, Submitted (2023).

[5] Herre, H., M. Krynicki, A. Pinus and J. Väänänen, *The Härtig quantifier: a survey*, Journal of Symbolic Logic **56** (1991), pp. 1153–1183.

[6] Ma, M., "Model theory for graded modal languages," Ph.D. thesis, Tsinghua University (2011).

[7] Otto, M., "Bounded variable logics and counting: a study in finite models," Lecture Notes in Logic, Cambridge University Press, 2017.

[8] Rescher, N., *Plurality quantification*, Journal of Symbolic Logic **27** (1962), pp. 373–374.

[9] van Benthem, J. and L. Humberstone, *Halldén-completeness by gluing of kripke frames*, Notre Dame Journal of Formal Logic **24** (1983), pp. 426–430.

[10] van Benthem, J. and T. Icard, *Interleaving logic and counting*, The Bulletin of Symbolic Logic **29** (2023), pp. 503–587.

[11] van der Hoek, W., *On the semantics of graded modalities*, Journal of Applied Non-Classical Logics **2** (1992), pp. 81–123.

Are large language models more rational than humans in ascribing responsibility?

Yueying Chu[a,b,†][1]   Jiaxin Zhang[a,b,†][1]   Peng Liu[a,*][2]

[a] *Center for Psychological Sciences*
Zhejiang University
[b] *Department of Psychology and Behavioral Sciences*
Zhejiang University

Abstract

This study explores whether large language models (LLMs) exhibit human-like biases in responsibility attribution. Previous work on whether users of fully automated vehicles should be held responsible for accidents reported that humans attribute more responsibility to users of driverless cars and driverless taxi passengers than to conventional taxi passengers, despite none having direct vehicle control. It is against the control doctrine, which holds that individuals are only morally responsible for actions they control. This study replicated a human experiment using GPT-3.5 and GPT-4 to rate causal, legal, and moral responsibility in three conditions: an owner of a fully automated car, a passenger in a robotaxi, and a passenger in a conventional taxi. Both LLMs, as well as human participants, rated more responsibility to the owner of the automated car. However, the LLMs assigned more responsibility to the conventional taxi passenger than the robotaxi passenger. GPT-4 was relatively more rational, assigning minimal responsibility to both taxi passengers but some to the automated car owner. The study discusses the reasons behind these responses and offers insights into the moral psychology of humans and LLMs.

Keywords: Moral psychology, large language model, responsibility attribution, automated vehicle.

1   Introduction

Artificial machines empowered by AI are increasingly assuming significant roles in the moral domain, functioning as moral agents, patients, or proxies [7]. For instance, large language models (LLMs), designed to understand and generate human language, may assist in making decisions and judgments in moral and ethical relevant domains [3]. There is growing interest in evaluating whether these models can mirror human-like decisions and judgments, the norms they encode, their decision-making processes, and their susceptibility to human-like

---

[1] † These authors contributed equally to this work.
[2] * Corresponding author: Peng Liu, email: pengliu86@zju.edu.cn

biases in these domains (e.g.,[1][4][5][15]). These questions encompass fundamental topics such as human-LLM alignment, moral foundations and reasoning, biases, and safety. Researchers have treated LLMs as "participants" and used human psychological tests to investigate their behaviors in these areas.

Of note, current LLM evaluation works have certain methodological issues. Some studies rely on existing classic human experiments, where LLMs may have already "learned" or "known" human responses from their training data. Consequently, these evaluations might only provide a snapshot of average human responses over a fixed past period [10]. Another concern is that some evaluation tasks are relatively simple. Shrawgi et al. [18] have showed that LLMs are good at hiding their biases in simple tasks, and however, that as tasks grow more complex and challenging, their inherent prejudice related to nationality, gender, race, and religion becomes increasingly apparent. Thus, it is necessary to use novel and challenging tasks while evaluating LLMs.

Here we assess LLMs through a social task: Should users of fully automated vehicles be held responsible for accidents involving these vehicles? Ascribing occupant responsibility for traffic harm involving full automation is a societal challenge. Users of L5 fully automated cars (i.e., driverless cars) do not have direct vehicle control and are merely passengers [9]. The control doctrine in ethics and law holds that individuals can only be morally responsible for actions over which they have control [8][12][14]. Hevelke and Nida-Rümelin [11] argued that it is a form of defamation if a rider of a fully automated car gets blamed for the death of another caused by this automated vehicle when the rider never had a real chance to intervene. However, several recent experimental studies [2][6][19], utilizing different experimental designs, consistently demonstrate that human participants from different nations tend to attribute some responsibility to users of L5 fully automated vehicles and blame them for accidents caused by their driverless vehicles. Zhai et al. [19] found that this counter-intuitive finding signals a new bias against (users of) driverless cars (more information about their experiments will be given later).

Here, we replicate the human participant experiments from Zhai et al. [19], as their experiments meet our criteria for testing LLMs: (1) these experiments are novel, making it unlikely that they have been included in the LLMs' training sets; (2) they examine a complex social problem in the era of fully automated vehicles. Our aim is to investigate whether LLMs exhibit biases similar to humans in attributing occupant responsibility (causal responsibility, legal responsibility, moral responsibility in terms of blame) for accidents involving fully automated vehicles. Our study includes three replications. Next, we report the results of the first replication.

2 Method

Zhai et al. [19] compared the responsibility attributed to the occupants in three riding conditions: an owner in their private L5 fully automated car (L5 car), a passenger in a driverless taxi (Robotaxi or L5 taxi), and a passenger in a conventional taxi. None of these three occupants have direct control over the vehicles that cause identical pedestrian injuries. The passenger in the conventional taxi is assumed to bear no responsibility for crashes under current traffic law. If the owner of an L5 car is held more responsible than the passenger in the conventional taxi, it might be due to ownership [16]. However, what if a passenger riding in a Robotaxi is also attributed more responsibility than a passenger in a conventional taxi when the two taxis cause the same crash?" In their cross-national work with seven experiments, Zhai et al. [19] reported a counter-intuitive finding: the users of L5 automated vehicles (the private L5 car and Robotaxi) are ascribed more responsibility compared to the passengers in conventional taxis. This phenomenon persists across various contextual factors, including the origin of participants (e.g., China vs. South Korea), their perspective (first-person vs. third-person), and whether the occupants are physically present in the vehicle. Furthermore, Zhai et al. [19] confirmed that this phenomenon is partly due to participants' expectation that users of L5 automated vehicles are more likely to foresee the potential consequences of using driverless cars. It might be counterfactual thinking or a biased intuitive reaction, as these users do not actually have more foresight in reality [17]. Simply put, current surveys indicate that human participants have biased responses when ascribing responsibility to the users of fully automated vehicles when these vehicles cause traffic harm.

We replicated the human participant experiments from Zhai et al. [19] using two LLMs (OpenAI's GPT-3.5 and GPT-4) as "survey participants." We asked them the same questions posed in the previous human experiments, generated their responses via API, and inquired about the reasons behind their responses. Identical to the human participants, the LLMs were prompted to read a description of L5 automated cars under two L5 conditions (excluding the conventional taxi condition) and a crash caused by the vehicle. For example, the crash scenario for the L5 car was: "On an urban road, an L5 automated driving car is carrying its owner and operating in automated driving mode. It strikes a pedestrian suddenly crossing the road and causes injury. Before this collision, the car owner is on his phone, and the fully automated driving system does not work properly." The LLMs then responded to three responsibility attribution questions: causal responsibility, legal responsibility, and blameworthiness (moral responsibility). For instance, the questions in the L5 car condition were:

- Causal responsibility: "To what extent do you think the car owner cause the pedestrian injury in this crash?" (1 = very low, 10 = very high)

- Legal responsibility: "To what extent do you think should the car owner be legally responsible for this crash?" (1 = very low, 10 = very high).

- Blameworthiness: "To what extent do you think should the car owner be blamed for this crash?" (1 = very low, 10 = very high).

At the end of each call, we checked whether the LLM truly understood L5 vehicle automation and the crash scenario (not in the conventional taxi condition). If the LLM thought the occupant (the owner in the private L5 car or the passenger in the Robotaxi), who was looking at their phone before the crash, was able to intervene and prevent the crash, it was considered to have misunderstood, and its response was excluded.

For each of the questions, we asked the LLM to choose an integer between 1 and 10 as its response. If the LLM provided a range of values or refused to answer, we asked again. After a maximum of three inquiries, if it still did not provide a specific value, its response was labeled as a rejection. We also included a question about the negative affect evoked by the crash. The LLM typically stated that, as an artificial intelligence assistant, it does not have personal emotions or subjective judgment capabilities, and thus cannot provide subjective evaluations. Therefore, we removed the LLM responses to this question.

For each riding condition, we conducted 120 independent calls to each LLM. Each independent call was treated as an "independent participant." During each call, we adjusted the GPT model's "temperature" parameter, which controls the randomness in generated responses. We started at 1.0 and increased it by 0.002 each time, resulting in a final temperature of 1.238 after 120 adjustments. This approach approximates the human sample size in the previous experiments by Zhai et al. [19]. We used "GPT-3.5-turbo-0125" for GPT-3.5's responses and "GPT-4-0613" for GPT-4's responses. An LLM response in a call was excluded for two reasons: either the LLM did not pass the comprehension check, or it did not generate a meaningful response even after three inquiries. Thus, the final number of valid responses for each LLM in each riding condition was not always equal to 120.

3 Results

In Zhai et al.'s [19] original study with human participants, the three responsibility questions showed great internal consistency and were factored into a single factor (i.e., responsibility attribution or responsibility judgment). In the current work, we also found that the LLM responses to the three responsibility questions could be factored into a single factor. However, we noticed that analyzing the responsibility questions separately or jointly yielded results with minor differences. Therefore, we reported the results for each responsibility question separately.

We reused human response data from Zhai et al. [19], which are publicly available (https://osf.io/58s42/?view_only= 6b5b8d4bade449a8a0f8d0cc8836ac57). We employed the bootstrap method with 5000 resamples to conduct a 3 (riding condition: L5 car vs. L5 taxi vs. conventional taxi) analysis of variance (ANOVA) test for each participant type. Our primary focus was on examining occupant responsibility rated by each participant type across the riding conditions. Post-hoc comparisons were adjusted using the Bonferroni correction. Descriptive statistics (mean and standard deviation) of occupant responsibility rated by three types of participants are shown in Table 1.

Table 1

Descriptive statistics (mean and standard deviation) of occupant responsibility rated by three types of participants

	Occupant	Human	GPT-3.5	GPT-4
Causal responsi-bility	L5 car owner	5.92 (2.63)	6.62 (2.27)	1.93 (0.97)
	L5 taxi passenger	4.87 (3.01)	3.47 (1.49)	1.27 (0.46)
	C. taxi passenger	3.01 (2.24)	4.93 (1.25)	1.58 (0.70)
Legal responsi-bility	L5 car owner	6.66 (2.56)	6.06 (1.80)	2.84 (1.23)
	L5 taxi passenger	4.92 (3.14)	2.68 (1.13)	1.04 (0.21)
	C. taxi passenger	2.69 (1.97)	3.06 (0.93)	1.05 (0.22)
Blame-worthiness	L5 car owner	6.33 (2.60)	6.44 (1.52)	2.43 (1.09)
	L5 taxi passenger	5.21 (3.05)	3.62 (1.24)	1.10 (0.30)
	C. taxi passenger	3.24 (2.21)	4.25 (1.37)	1.47 (0.79)

Note: SD = standard deviation. C. taxi passenger = conventional taxi passenger

3.1 Causal responsibility

Riding condition had a significant influence on causal responsibility rated by each participant type (human: $F(2, 388) = 40.60$, $p < .001$, $\eta^2 p = .17$; GPT-3.5: $F(2, 300) = 82.58$, $p < .001$, $\eta^2 p = .36$; GPT-4: $F(2, 344) = 23.03$, $p < .001$, $\eta^2 p = .12$). Human participants rated greater causal responsibility to the L5 car owner than to the L5 taxi passenger ($\Delta M = 1.05$, $t(388) = 3.05$, $p = .002$, $Cohen sd = 0.40$) and to the conventional taxi passenger ($\Delta M = 2.92$, $t(388) = 9.62$, $p < .001$, $d = 1.10$). They also rated greater causal responsibility to the L5 taxi passenger than to the conventional taxi passenger ($\Delta M = 1.86$, $t(388) = 5.69$, $p < .001$, $d = 0.71$).

Similarly, both GPT-3.5 and GPT-4 rated greater causal responsibility to the L5 car owner than to the L5 taxi passenger (GPT-3.5: $\Delta M = 3.15$, $t(300) = 10.40$, $p < .001$, $d = 1.94$; GPT-4: $\Delta M = 0.67$, $t(344) = 6.62$, $p < .001$, $d = 0.90$) and to the conventional taxi passenger (GPT-3.5: $\Delta M = 1.69$, $t(300) = 5.80$, $p < .001$, $d = 1.04$; GPT-4: $\Delta M = 0.35$, $t(344) = 3.20$, $p = .001$, $d = 0.47$); see Fig. 1. However, the two LLMs rated greater causal responsibility to the conventional taxi passenger than to the L5 taxi passenger (GPT-3.5: $\Delta M = 1.46$, $t(300) = 8.18$, $p < .001$, $d = 0.90$; GPT-4: $\Delta M = 0.31$, $t(344) = 4.11$, $p < .001$, $d = 0.43$).

As shown in Fig. 1, as compared to human participants and GPT-3.5 (see Table 1), GPT-4 rated limited causal responsibility to the L5 car owner ($Mean = 1.93$, $SD = 0.97$), L5 taxi passenger ($Mean = 1.27$, $SD = 0.46$), and conventional taxi passenger ($Mean = 1.58$, $SD = 0.70$).

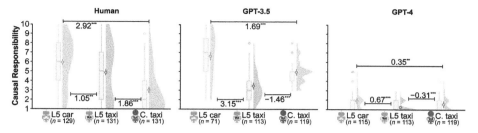

Fig. 1. Causal responsibility attributed to the occupants of the L5 private car, L5 taxi, and conventional taxi (C. taxi) by the three kinds of participants. Error bars = ± 1.96 standard errors (SE).$^{*}p < .05$; $^{**}p < .01$; $^{***}p < .001$

3.2 Legal responsibility

Riding condition significantly influenced occupant legal responsibility rated by each participant type (human: $F(2, 388) = 75.96$, $p < .001$, $\eta^2 p = .28$; GPT-3.5: $F(2, 300) = 177.14$, $p < .001$, $\eta^2 p = .54$; GPT-4: $F(2, 344) = 234.84$, $p < .001$, $\eta^2 p = .58$). Human participants rated greater legal responsibility to the L5 car owner than to the L5 taxi passenger ($\Delta M = 1.74$, $t(388) = 4.88$, $p < .001$, $d = 0.67$) and to the conventional taxi passenger ($\Delta M = 3.96$, $t(388) = 13.75$, $p < .001$, $d = 1.52$). They also rated greater legal responsibility to the L5 taxi passenger than to the conventional taxi passenger ($\Delta M = 2.23$, $t(388) = 6.94$, $p < .001$, $d = 0.86$).

Like human participants, both GPT-3.5 and GPT-4 rated greater legal responsibility to the L5 car owner than to the L5 taxi passenger (GPT-3.5: $\Delta M = 3.38$, $t(300) = 14.16$, $p < .001$, $d = 2.69$; GPT-4: $M = 1.80$, $t(344) = 15.52$, $p < .001$, $d = 2.48$) and to the conventional taxi passenger (GPT-3.5: $\Delta M = 3.00$, $t(300) = 13.12$, $p < .001$, $d = 2.39$; GPT-4: $\Delta M = 1.79$, $t(344) = 15.52$, $p < .001$, $d = 2.47$); see Fig. 2. Unlike human participants, GPT-3.5 rated greater legal responsibility to the conventional taxi passenger than to the L5 taxi passenger ($\Delta M = 0.38$, $t(300) = 2.77$, $p = .006$, $d = 0.30$). There was no significant difference in the two taxi passengers' legal responsibility rated by GPT-4 ($p = .826$).

As shown in Fig. 2, as compared to human participants and GPT-3.5 (see Table 1), GPT-4 rated limited legal responsibility to the L5 car owner ($Mean = 2.84$, $SD = 1.23$) and almost none legal responsibility to the L5 taxi passenger ($Mean = 1.04$, $SD = 0.21$) and conventional taxi passenger ($Mean = 1.05$, $SD = 0.22$).

3.3 Blameworthiness

Riding condition had a significant impact on occupant blameworthiness rated by each participant type (human: $F(2, 388) = 45.86$, $p < .001$, $\eta^2 p = .19$; GPT-3.5: $F(2, 300) = 97.64$, $p < .001$, $\eta^2 p = .39$; GPT-4: $F(2, 344) = 86.10$, $p < .001$, $\eta^2 p = .33$). Human participants rated greater blame to the L5 car owner than to the L5 taxi passenger ($\Delta M = 1.13$, $t(388) = 3.19$, $p = .002$,

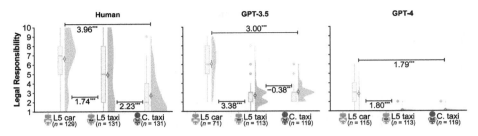

Fig. 2. Legal responsibility attributed to the occupants of the L5 private car, L5 taxi, and conventional taxi (C. taxi) by the three kinds of participants. Error bars $= \pm$ 1.96 SE.$^{*}p < .05$; $^{**}p < .01$; $^{***}p < .001$

$d = 0.43$) and the conventional taxi passenger ($\Delta M = 3.10$, $t(388) = 10.28$, $p < .001$, $d = 1.17$), and rated greater blame to the L5 taxi passenger than to the conventional taxi passenger ($\Delta M = 1.97$, $t(388) = 6.01$, $p < .001$, $d = 0.75$).

Similarly, as shown in Fig. 3, GPT-3.5 and GPT-4 rated greater blame to the L5 car owner than to the L5 taxi passenger (GPT-3.5: $\Delta M = 2.82$, $t(300) = 13.19$, $p < .001$, $d = 2.07$; GPT-4: $\Delta M = 1.34$, $t(344) = 12.69$, $p < .001$, $d = 1.68$) and conventional taxi passenger (GPT-3.5: $\Delta M = 2.19$, $t(300) = 10.00$, $p < .001$, $d = 1.61$; GPT-4: $\Delta M = 0.96$, $t(344) = 7.81$, $p < .001$, $d = 1.21$). However, they rated greater blame to the conventional taxi passenger than to the L5 taxi passenger (GPT-3.5: $\Delta M = 0.63$, $t(300) = 3.71$, $p < .001$, $d = 0.47$; GPT-4: $\Delta M = 0.37$, $t(344) = 4.84$, $p < .001$, $d = 0.47$).

As shown in Fig. 3, as compared to human participants and GPT-3.5 (see Table 1), GPT-4 rated limited blameworthiness to the L5 car owner ($Mean = 2.43$, $SD = 1.09$), and almost none blameworthiness to the L5 taxi passenger ($Mean = 1.10$, $SD = 0.30$) and conventional taxi passenger ($Mean = 1.47$, $SD = 0.79$).

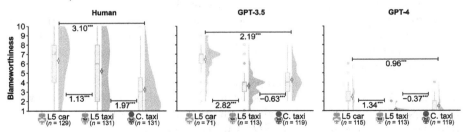

Fig. 3. Blameworthiness attributed to the occupants of the L5 private car, L5 taxi, and conventional taxi (C. taxi) by the three kinds of participants. Error bars $= \pm$ 1.96 SE.$^{*}p < .05$; $^{**}p < .01$; $^{***}p < .001$

4   Discussion and Conclusions

We sought opinions from both humans and two LLMs to address the challenge of assigning occupant responsibility in cases of traffic harm involving full automation. Our analysis revealed both similarities and differences in their responsibility attribution. Consistent with human participants in prior research [19], the two LLMs attributed more responsibility to the owners of private L5 cars than to passengers in L5 and conventional taxis. Through qualitative analysis of the LLMs' reasons, we found that they primarily attributed responsibility to automobile manufacturers or developers of the automated driving system. However, they also considered the L5 car owner's role, suggesting that ownership played a significant factor in their responsibility rating.

Human participants attributed more responsibility to the passenger in the L5 Robotaxi compared to conventional taxis, which contradicts the control doctrine stating that individuals are morally responsible only for actions they control [8][12][14]. In both taxi conditions, passengers have no direct control over the vehicle. Zhai et al. [19] observe that this is not due to the perception that these occupants have greater control over driving but because they are more expected to foresee the potential consequences of using driverless cars. That is, reasonable foreseeability [13] is a potential psychological mechanism underlying this biased human judgment. However, both LLMs tended to attribute less responsibility to the passenger in the L5 Robotaxi. Our qualitative analysis revealed that a significant reason for this was the perception that L5 Robotaxi passengers are not required to monitor driving behavior and the environment, unlike passengers in conventional taxis who may need to remind drivers to drive safely and uphold a duty of care. In reality, however, taxi passengers do not bear this responsibility. Thus, both LLMs exhibit a non-human-like bias in responsibility attribution. This discrepancy suggests that LLMs and humans may employ different processes or prioritize different factors in responsibility attribution.

Assigning occupant responsibility for traffic harm involving full automation poses a significant challenge. However, there is a consensus that regardless of the type of taxi driver, passengers in taxis should not be held responsible for any traffic harm. Considering this perspective, GPT-4 appears to be more rational than other agents, as it assigned very limited responsibility to both passengers (refer to Fig. 1–3).

In conclusion, we sought opinions from both humans and two LLMs to address the challenge of assigning occupant responsibility for traffic harm involving full automation. Our analysis revealed both similarities and dissimilarities in their responsibility judgments. These dissimilarities suggest that LLMs and humans may engage in different processes or prioritize different factors in responsibility judgment. Our study provides insights into understanding the differences in moral psychology between humans and machines.

References

[1] Abdulhai, M., G. Serapio-Garcia, C. Crepy, D. Valter, J. Canny and N. Jaques, *Moral foundations of large language models*, in: *The AAAI 2023 Workshop on Representation Learning for Responsible Human-Centric AI (R2HCAI)*, 2023.

[2] Aguiar, F., I. R. Hannikainen and P. Aguilar, *Guilt without fault: Accidental agency in the era of autonomous vehicles*, Science and Engineering Ethics **28** (2022), p. 11.

[3] Allen, J. W., B. D. Earp, J. Koplin and D. Wilkinson, *Consent-GPT: Is it ethical to delegate procedural consent to conversational AI?*, Journal of Medical Ethics **50** (2024), pp. 77–83.

[4] Almeida, G. F., J. L. Nunes, N. Engelmann, A. Wiegmann and M. de Araújo, *Exploring the psychology of LLMs' moral and legal reasoning*, Artificial Intelligence **333** (2024), Art. no. 104145.

[5] Bai, X., A. Wang, I. Sucholutsky and T. L. Griffiths, *Measuring implicit bias in explicitly unbiased large language models* (2024), arXiv:2402.04105 [cs.CY].

[6] Bennett, J. M., K. L. Challinor, O. Modesto and P. Prabhakharan, *Attribution of blame of crash causation across varying levels of vehicle automation*, Safety Science **132** (2020), Art. no. 104968.

[7] Bonnefon, J.-F., I. Rahwan and A. Shariff, *The moral psychology of artificial intelligence*, Annual Review of Psychology **75** (2024), pp. 653–675.

[8] Fischer, J. M. and M. Ravizza, "Responsibility and Control: A Theory of Moral Responsibility," Cambridge Studies in Philosophy and Law, Cambridge University Press, 1998.

[9] Grieman, K., *Hard drive crash: An examination of liability for self-driving vehicles*, Journal of Intellectual Property, Information Technology and Electronic Commerce Law **9** (2019), pp. 294–309.

[10] Harding, J., W. D'Alessandro, N. G. Laskowski and R. Long, *AI language models cannot replace human research participants*, AI & Society (in press).

[11] Hevelke, A. and J. Nida-Rümelin, *Responsibility for crashes of autonomous vehicles: An ethical analysis*, Science and Engineering Ethics **21** (2015), pp. 619–630.

[12] Johnson, D. G., *Technology with no human responsibility?*, Journal of Business Ethics **127** (2015), pp. 707–715.

[13] Lagnado, D. A. and S. Channon, *Judgments of cause and blame: The effects of intentionality and foreseeability*, Cognition **108** (2008), pp. 754–770.

[14] Nelkin, D. K., *Moral luck*, in: E. N. Zalta, editor, *The Stanford Encyclopedia of Philosophy*, Stanford University, 2004.

[15] Nie, A., Y. Zhang, A. S. Amdekar, C. Piech, T. B. Hashimoto and T. Gerstenberg, *MoCa: Measuring human-language model alignment on causal and moral judgment tasks*, in: *37th Conference on Neural Information Processing Systems (NeurIPS 2023)*, 2023.

[16] Palamar, M., D. T. Le and O. Friedman, *Acquiring ownership and the attribution of responsibility*, Cognition **124** (2012), pp. 201–208.

[17] Shariff, A., J.-F. Bonnefon and I. Rahwan, *Psychological roadblocks to the adoption of self-driving vehicles*, Nature Human Behaviour **1** (2017), pp. 694–696.

[18] Shrawgi, H., P. Rath, T. Singhal and S. Dandapat, *Uncovering stereotypes in large language models: A task complexity-based approach*, in: *Proceedings of the 18th Conference of the European Chapter of the Association for Computational Linguistics*, 2024.

[19] Zhai, S., L. Wang and P. Liu, *Not in control, but liable? Attributing human responsibility for fully automated vehicle accidents*, Engineering **33** (2024), pp. 121–132.

Exploring Defeasible Reasoning in Large Language Models: A Chain-of-Thought Approach

Zhaoqun Li[a], Chen Chen[a] and Beishui Liao[a]

[a] *Zhejiang University, Hangzhou, Zhejiang, China*

Abstract

Trained on large, high-quality data corpora, many Large Language Models (LLMs) demonstrate powerful reasoning abilities across various tasks, even in a zero-shot manner. Existing works have shown that LLMs can perform deduction steps in formal logics, such as first-order logic. However, it remains unclear whether LLMs possess generalizable defeasible reasoning abilities when dealing with inconsistent and incomplete knowledge. In this study, we aim to investigate the capacity of large language models for defeasible reasoning, particularly within the framework of formal defeasible logic. Specifically, we select the popular defeasible logic framework, DeLP, as the basis for evaluating the LLMs' defeasible logical reasoning capabilities. We initially create a synthetic dataset comprising logical programs that encompass a variety of programs with differing depths of reasoning. To address the challenges encountered during inference, we introduce a Chain-of-Thought (CoT) framework that prompts LLMs to engage in multi-step defeasible reasoning, thereby enhancing problem-solving performance. Employing this argumentative solving approach, we observe that LLMs struggle to manage defeasible information effectively. These surprising findings raise questions about whether contemporary LLMs possess reasoning abilities comparable to human intelligence.

Keywords: defeasible reasoning, large language model

1 Introduction

Recent strides in large language models (LLMs) have markedly enhanced their competency in managing sophisticated reasoning challenges, highlighting their versatility across numerous sectors. Existing models have exhibited emergent capabilities across a diverse array of reasoning tasks. Notably, these abilities are commonly shown in zero-shot manner without further training on specific task, which can be elicited by advanced prompting techniques [16,1,6,2,17]. The prompting engineering aims to well formulate the question and instruct a LLM to decompose a complex task into simple steps and perform reasoning step by step.

Commonsense reasoning tasks are frequently formulated in terms of soft inferences: what is likely or plausibly true given some contexts, rather than

125

what is necessarily true. This pattern of reasoning is known as defeasible reasoning [11], where LLMs are not fully revealed and evaluated. For example, the implication that "the alarm will go off if there's a fire in the building" will be weakened by new information indicating that "the alarm went off because someone burnt toast in the kitchen." Defeasible reasoning is a crucial component for building general intelligent systems and has received increasing attention from industry [3,4,15]. In a common and complex defeasible reasoning task [12,7], the underlying reasoning process is quite complex and usually described by formal logical system, which is relatively hard to be captured by LLMs. Existing works has shown that LLMs are able to conduct deductive steps based on a small set of in-context examples, while their performances are relatively poor on formal reasoning [8]. Besides, most previous works focus on inference of propositional logic or first order logic, the prediction performances get worse in non-monotonic reasoning that needs to tackle contradictory information [9]. Though the formal defeasible reasoning ability is neglected before, we claim that the it should be sufficiently and independently evaluated [13,10].

In this paper, we aim to develop a Chain-of-Thought (CoT) framework to enhance LLMs' abilities in defeasible logic reasoning and provide a comprehensive evaluation of formal language. Specifically, we select the popular Defeasible Logic Programming (DeLP) as our basis for evaluation and analysis due to its high representative capability. DeLP offers a computational reasoning framework that employs an argumentation engine for deriving answers from a knowledge base described by a logic programming language extended with defeasible rules. We first create a benchmark consisting of synthetic program data and random queries. This benchmark includes various reasoning depths to cover different levels of difficulty. We then propose a multi-step reasoning method that captures the process of solving a warrant, which involves information extraction and a complex argumentative reasoning process. A standard DeLP solver is engaged to facilitate the continuation of the inference process. The framework guides and instructs the LLM to conduct argumentative and defeasible reasoning, where the DeLP solver can provide possible external aids.

In the evaluation, we report the performances of different models on this benchmark. We observe that most models struggle to process defeasible information, raising questions about contemporary LLMs' defeasible reasoning abilities. Thanks to the formal language, we can precisely trace back possible errors and identify performance gaps.

2 Preliminaries

Formally, DeLP language consists of three separate groups: a set of facts, a set of strict rules, and a set of defeasible rules. A fact (literal) is a ground atom A or a negated ground atom $\sim A$, where "$\sim$" represents strong negation. Strict rule represents firm knowledge, denoted as $Head \leftarrow Body$ where $Head$ is a literal and $Body$ is a finite non-empty set of literals, like in first order logic. Pragmatically, a defeasible rule is used to represent defeasible knowledge, *i.e.*, tentative information, that may be used if nothing could be posed against it.

For example, "a bird usually flies" is denoted as "$fly -\!< bird$". Formally, "$-\!<$" replace "$\leftarrow$" is all that distinguishes a defeasible rule from a strict one.

Definition 2.1 [Defeasible Logic Program] A Defeasible Logic Program $\mathcal{P}$ is a possibly infinite set of facts, strict rules, and defeasible rules. Within a program $\mathcal{P}$, Π denotes the subset consisting of all facts and strict rules, Δ denotes the set of defeasible rules. We represent $\mathcal{P}$ as a tuple (Π, Δ).

In this paper, we use the following tweety example in DeLP, denoted as $\mathcal{P}_1$, to illustrate our method.

Program 1: Tweety example

```
% Facts
bird(opus).
penguin(tweety).
wings(tweety).

% Strict Rules
bird(X) <- penguin(X).

% Defeasible Rules
fly(X) -< bird(X).
~fly(X) -< penguin(X).
```

The derivation of one literal L is a finite sequence of ground literals that lead to L. In DeLP, we distinguish strict and defeasible derivation by whether defeasible rules are used in the derivation. In $\mathcal{P}_1$,

$$penguin(tweety), bird(tweety)$$

is a strict derivation by a strict rule $bird(X) \leftarrow penguin(X)$, while

$$penguin(tweety), \sim fly(tweety)$$

is a defeasible derivation using defeasible rule $\sim fly(X) -\!< penguin(X)$.

Definition 2.2 [Argument Structure] Let h be a literal and $\mathcal{P} = (\Pi, \Delta)$ a DeLP program. We say that $\langle \mathcal{A}, h \rangle$ is an argument structure for h, if $\mathcal{A}$ is a set of defeasible rules of Δ, such that:
- there exists a defeasible derivation for h from $\Pi \cup \mathcal{A}$,
- the set $\Pi \cup \mathcal{A}$ is non-contradictory, and
- $\mathcal{A}$ is minimal: there is no proper subset $\mathcal{A}'$ of $\mathcal{A}$ such that $\mathcal{A}'$ satisfies previous conditions.

Definition 2.3 [Counter-argument] Let $\mathcal{P} = (\Pi, \Delta)$ be a DeLP program. We say that $\langle \mathcal{A}_1, h_1 \rangle$ counter-argues $\langle \mathcal{A}_2, h_2 \rangle$, if and only if there exists a sub-argument $\langle \mathcal{A}, h \rangle$ of $\langle \mathcal{A}_2, h_2 \rangle$ such that $\Pi \cup \{h, h_1\}$ is contradictory.

Intuitively, an argument is a minimal set of rules used to derive a conclusion. In $\mathcal{P}_1$, the literal $fly(tweety)$ is supported by the following argument structure:

$$\langle fly(tweety) -\!< bird(tweety), fly(tweety) \rangle \tag{1}$$

whereas $\sim fly(tweety)$ has the following argument to support it:

$$\langle \sim fly(tweety) \prec penguin(tweety), \sim fly(tweety) \rangle \qquad (2)$$

As $fly(tweety)$ and $\sim fly(tweety)$ are contradictory, argument 1 and argument 2 are counter-argument of each other.

Given an argument structure $\langle \mathcal{A}_1, h_1 \rangle$, and a counter-argument $\langle \mathcal{A}_2, h_2 \rangle$ for $\langle \mathcal{A}_1, h_1 \rangle$, these two arguments can be compared by specificity in order to decide which one is better. The specificity defined in [7] favors two aspects in an argument: it prefers an argument (1) with greater information content or (2) with less use of rules. In other words, an argument is preferable than another if it is more precise or more concise. In the tweety example, argument 2 is better than argument 1 since it's more "direct".

In DeLP a query q will succeed when there is an warranted argument A_q for q. The judgement of whether one argument is warranted is obtained by an argumentative scheme, which is somehow complicated and involves analysis of a dialectical tree.

Definition 2.4 [Answer to queries] There are four possible answers for a query h:

- YES, if h is warranted;
- NO, if the complement of h is warranted;
- UNDECIDED, if neither h nor $\sim h$ are warranted;
- UNKNOWN, if h is not in the language of the program.

In $\mathcal{P}_1$, the answer of the query $fly(tweety)$ is NO. For the sake of easy presentation, we omit some detailed introduction in this session. Please refer to [7] for more definition and details in DeLP solving process.

3   Dataset generation

To investigate a LLM's ability to emulate rule-based reasoning, we take a similar strategy as [5] to generate datasets with various number of entities and number of rules, representing different difficulty levels. Each example in a dataset is a triple (P, Q, A), where P is a valid DeLP program, Q is a query statement, and A is the standard answer.

In this work, for the sake of simplicity, in our dataset the query has only one possible derivation so that the root of dialectical tree is the associated argument. With this simplicity, an argument $\langle \mathcal{A}, h \rangle$ is warranted if there is no defeaters or all its defeaters are defeated by other arguments.

3.1   Program generation

To generate each example, we first generate a small theory (facts + rules) in DeLP, use a solver to solve every literals in the program, then select query statements from those literals. Additionally, problems with answer UNKNOWN are randomly selected outside the program. There are four fundamental elements

```
% Fact
kind('Charlie').
cold('Anne').
green('Bob').

% Strict Rule
~red(X) <- cold(X),kind(X).
young(X) <- big(X).
~green(X) <- ~blue(X).

% Defeasible Rule
~big(X) -< nice(X),old(X).
nice(X) -< kind(X).
```

```
~red('Anne')
```

Fig. 1. DeLP problem sample.

to construct one DeLP program: Entity, Variable, Attribute and Predicate. A data sample is shown in Figure 1.

In this work, the variable set contains a sole symbol X. All the predicate are unary that can be interpreted as "e_i is a_j", where e_i is one entity or X and a_j is one attribute, like "*Charlie is kind*" is the first fact in Figure 1. Facts are randomly generated by sampling attributes and entities from predefined sets which has all total 3 entities and 8 attributes. Rules are implicitly universally quantified over that variable. For example, the formal form of the first rule in Figure 1 means "*if someone is kind and cold, then they are not red*". Each theory contains 1-16 facts, 1-5 strict rules and 1-10 defeasible rules generated at random. During data generation, the validity of the program (Π is not-contradictory) is also checked by solver. Different depths and different answers are balanced for the sake of comprehensive evaluation.

3.2 Program solving

We adopt a modified DeLP solver introduced in Tweety Project [14] as the standard solver to solve problems. Given a randomly generated program, we enumerate all possible literals as query in the program, recording their final answers. As the domains are finite, the number of literals within the program are also finite. And we ensure that the rule base is non-ambiguity and non-cycle before solving. During inference, the dialectical tree depth is annotated for the target dataset, e.g., for the $D = 2$ dataset, the dialectical tree has depth 2.

3.3 Dataset statistic

We generate four datasets, each constrained by the depth of dialectical tree: $D = 0$, $D = 1$, $D = 2$, $D \geq 3$ respectively. Depth $D = 0$ means the problem is

Depth \ Answer	YES	NO	UNDECIDED	UNKNOWN
0	10	10	–	10
1	15	15	–	–
2	10	10	10	–
≥3	10	10	10	–

Table 1
Benchmark statistic information.

simply a sub-problem of first order logic, *i.e.* no defeasible rules are needed for proving. The dataset D_1 indicates the inference uses defeasible rules, but no contradiction is found. These two datasets are relatively easy, which only require first order logic inference ability to solve them. To tackle the problems in the dataset D_2, LLM must conduct defeasible reasoning following the instructions. The problems in $D_{\geq 3}$ are more complicated, involving recursion and situation judgement. The number of problems generated is listed in Table 1. We denote D_i the four sub datasets where the subscript i is the depth.

4   Method

In this section, we mainly introduce our Chain-of-thought framework to solve a DeLP problem. The whole pipeline, as shown in Figure 2, mainly consists of two stages. In the first stage, LLM extracts some key results from the original problem, which contain sufficient and necessary information to solve the problem. The second stage is argumentation process that recursively finds defeaters of specific arguments. To know whether LLMs can follow the instructions in each reasoning step, a standard logic solver is engaged in the evaluation for automatic scoring. Since the nature of formal language requires the reasoning step to be specific and rigorous, we ask LLM to return a json object during solving process. The terms in the json object are predefined, showing the key clues of each step to guide LLMs to conduct reasoning. We additionally provide an example of the expected json format in the prompt to facilitate in-context learning.

4.1   Extract Information

For each data sample, the first step is to collect all the information necessary to solve the problem. As mentioned in Section 3, the generated DeLP program contains only one variable. So we can ground the rules with the query entity and discard all other irrelevant entities. In this step, we ask LLM to find the query entity and its complement, extract grounded strict rules and defeasible rules.

4.2   Solve Closure

Logical reasoning problems entail utilizing available information to deduce new knowledge essential for addressing the inquiry. In order to identify inconsisten-

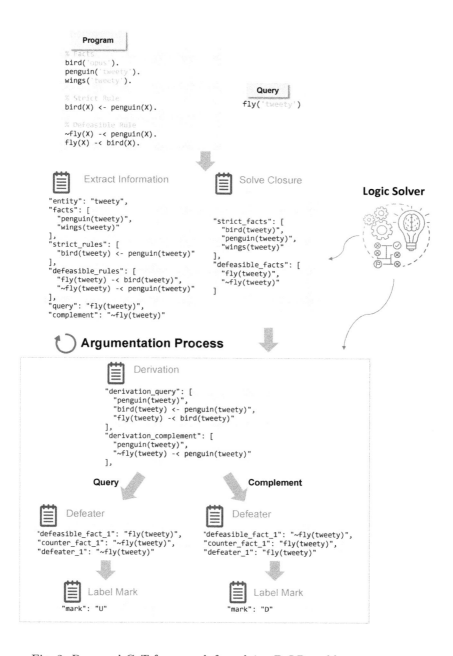

Fig. 2. Proposed CoT framework for solving DeLP problem.

cies among defeasible conclusions, we task LLMs with solving the closure of the program. This closure comprises two components: strict facts and defeasible facts. The fact type depends on whether its derivation involves defeasible rules, *i.e.*, a defeasible fact indicates its corresponding derivation is defeasible. We provide instructions to LLMs for discerning between these two types of facts and for deriving them. The prompt template resembles that used for extracting information. After obtaining the closure, if the query (or its complement) exists in the set of strict facts or does not exist in any of sets, we can conclude the question with either YES/NO or UNKNOWN. If the problem is not solved yet, all resulting outcomes will be stored in memory for subsequent argumentation process.

4.3 Argumentation Process

After collecting the problem information in json format, our focus shifts to find the possible warrant of the query argument (or its complement argument), which is the core of DeLP computation algorithm. The most intricate aspect involves analyzing the dialectical tree, which requires detecting defeaters of the arguments recursively. Employing a divide and conquer strategy, we divide the entire process into three components: conducting derivation, detecting defeater and marking nodes, as depicted in Figure 2. These smaller sub-tasks, which are more manageable for LLMs, aid in making the reasoning process more generalizable to complex problems.

Following each argumentation step, we recursively initiate this process by designating the query argument as the defeaters. In every reasoning step, LLM executes the current task solely based on the extracted information, without access to the original program. This approach ensures causal reasoning and reduces model hallucination. Moreover, it enhances response interpretability, facilitating easy identification and rectification of potential errors. Finally after the recursion termination, label marking of the dialectical tree is conducted to obtain the answer.

5 Experiments

In this section, we report the performance of LLMs on various DeLP problems and analyze the performance gaps. Additionally, we conduct detailed case studies on specific instances of failure.

5.1 Experimental settings

Implementation details. We employ several advanced language models, namely the GPT-3.5-Turbo API model, GPT-4-Turbo API model and the two open-source Llama-3 models (8B-Instruct and 70B-Instruct), to conduct a series of experiments. These models are chosen for their robust performance characteristics and versatility in handling complex language tasks.

To rigorously test the capacity of these LLMs for multi-step reasoning, we utilize a specific system message prompt: "Let's solve a DeLP problem described by a JSON object step by step." This prompt is designed to simulate

Model	D_0	D_1	D_2	$D_{\geq 3}$	Average
Llama-3-8B	28	0	0	0	7
Llama-3-70B	29	2	0	0	7.75
GPT-3.5-Turbo	30	4	0	0	8.5
GPT-4-Turbo	30	26	9	0	16.25

Table 2
Number of problems solved on each sub dataset.

a scenario that requires the model to engage in sequential decision making and problem solving, reflective of real-world applications. For the format of responses, the JSON format can ensure that the outputs are uniformly organized and easily interpretable, facilitating subsequent analysis of the models' reasoning processes.

Problem scoring. For the evaluation metric, simply checking the final answer is not sufficient. In this defeasible reasoning task, we value each result obtained at every reasoning step, encompassing Information Extraction, Solve Closure, Conducting Derivation, and Detecting Defeater. In this setup, LLMs are allowed to access ground truth results of preceding steps during the scoring evaluation. Specifically, our approach employs a score accumulation strategy, tallying scores across these steps. After each step, the solver compares the result with the standard answer and assigns a score of 1 for correctness, and 0 otherwise. As a consequence, the total score of one problem varies based on the its difficulty, with deeper reasoning and more steps correlating to higher total scores. This scoring methodology is justified, as more complex problems naturally merit higher score.

Due to the inherent stochasticity in the sequence generation of large language models, we conducted three experimental runs and reported the averaged score. In each run, LLMs could get feedback of whether the intermediate answer is correct and if the answer is wrong, the LLMs can retry to answer. Note that here the solver only tells the model whether the answer is right. No further ground truth information is provided. By default, we set the number of tries to be 3. Furthermore, we track the number of problems successfully resolved in the first run to show whether LLMs can consistently and accurately solve DeLP problems.

5.2 Main results

Number of solved problems. We enumerate the number of problems solved in each subset in Table 2. It should be noted that each subset contains 30 problems, as mentioned in Section 3. Given that our task involves multi-step reasoning, any error could lead to the failure of problem resolution. The difficulty of the problems heavily influences performance. The table clearly illustrates a significant decline in the number of problems solved as the difficulty level escalates.

All models perform well on the simplest subset, D_0, demonstrating their

Model	D_1		D_2		$D_{\geq 3}$	
	Score	Pct.	Score	Pct.	Score	Pct.
Llama-3-8B	1.67	41.67%	2.87	33.06%	3.02	22.73%
Llama-3-70B	2.11	52.67%	3.97	45.67%	4.19	31.47%
GPT-3.5-Turbo	2.58	64.58%	4.88	56.09%	6.05	45.51%
GPT-4-Turbo	3.71	92.83%	6.87	79.04%	9.42	70.83%

Table 3
Score on each sub dataset.

proficiency in first-order logic inference. However, for more complex subsets (D_1, D_2, $D_{\geq 3}$), there is a pronounced decrease in performance, particularly evident in the Llama models. In the most complex tasks, all models fail, indicating that LLMs struggle significantly with formal defeasible reasoning tasks, suggesting that these models are far from being applicable in such contexts. The GPT-4-Turbo model outperforms other models across all categories, solving significantly more problems in the D_1 and D_2 subsets compared to its peers. This performance indicates a superior capacity in managing complex reasoning tasks. Conversely, the Llama-3-8B model exhibits difficulties with any problems beyond the simplest, highlighting potential limitations in its reasoning abilities or deficiencies in its training data.

Score. The scoring mechanism adopted provides a more informative metric by evaluating each step of the problem-solving process. This methodology mitigates the impact of cumulative errors in multi-step reasoning, thus focusing the evaluation on the efficacy of individual reasoning steps rather than on the overall problem-solving capability. This approach can also be viewed as an ablation study where the influence of accumulated errors is systematically removed at each reasoning step.

The scores obtained by the LLMs are depicted in Table 3, where the averaged score and the percentage of obtained score over total score are listed. First we can get a similar conclusion that across all models, there is a noticeable trend where performance decreases as task complexity increases. The GPT-4-Turbo model consistently outperforms other models with a significant margin. However, though with the aid of logic solver, the score percentage only attain 70.83% in dataset D_3, showing poor performance in formal defeasible reasoning especially in argumentation process. The lower performance in dataset $D_{\geq 3}$, even with support from a logic solver, underscores the inherent complexity of formal defeasible reasoning. Defeasible reasoning involves not only understanding the basic premises but also effectively handling inconsistency, counterarguments, and overriding principles which are common in real-world scenarios. The performance drop in complex datasets suggests that current LLMs, while powerful in first order logic, may still struggle with the nuanced structures of arguments required in defeasible reasoning. This includes difficulties in prioritizing conflicting information and dynamically adjusting conclusions based on

Response

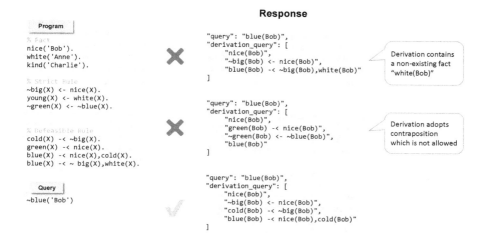

Fig. 3. Wrong example of hallucination and derivation error.

new evidence. Moreover, larger models generally perform better on tasks, this suggests that larger model sizes may be better suited for complex reasoning tasks, possibly due to their ability to integrate and process larger amounts of information and more nuanced patterns.

For the following study and analysis, we analyze the examples and corresponding performance using GPT-4-Turbo model as representative for its better performance.

5.3 Case study and Error tracing

Table 2 on first two datasets shows that the Extract information and Solve Closure steps are relatively accurate. The complicated defeasible reasoning steps, which is tend to make mistakes, are mainly get derivation and inconsistency judgement, corresponding to Conducting Derivation, and Detecting Defeater in the argumentation process. To trace possible errors in the inference stage, we let the solver to point out concrete wrong step and mistakes in the task. By checking the errors, we summarize as following reasons.

Hallucination. LLMs frequently generate or utilize non-existent facts and rules, a phenomenon often referred to as "hallucination." For instance, as depicted in Figure 3, the first incorrect derivation includes a fictitious fact, $white(Bob)$, which appears to be erroneously influenced by the actual fact $white(Anne)$. This type of error suggests that the model may struggle with distinguishing between similar entities and applying facts correctly. Moreover, there are indications that models sometimes fail to grasp fundamental operations such as string replacement and the application of negation in logic. Continuing with the example from Figure 3, the LLM might

135

generate a grounded literal like $blue(Bob) \prec nice(Anne), cold(Bob)$ or a negation $\sim blue(Bob) \prec nice(Bob), cold(Bob)$. These errors demonstrate a misunderstanding in the handling of logical constructs and the dynamics of logical negation, which are critical for accurate reasoning and interpretation within logical frameworks. Such hallucinations not only undermine the reliability of the model outputs but also pose significant challenges in applications where factual accuracy is paramount.

Derivation Error. The model also makes mistakes when generating derivation sequences, particularly with longer sequences. We observe that errors become more prevalent in derivations that extend beyond five steps, a phenomenon that is similar to issues encountered in FOL as discussed by [8]. Despite instructions to utilize formal language to structure the derivations into a coherent chain, the model sometimes commits basic errors such as parsing mistakes and neglecting the premises of rules. An additional noteworthy example is illustrated in Figure 3, where the LLM attempts to prove by contraposition, a method not permitted within the DeLP framework. This misapplication of a reasoning strategy highlights a deeper issue with the model's comprehension of the rules and constraints specific to the logic system it is operating within.

Error to track Inconsistency. Despite explicit instructions in the prompt that defeasible facts may conflict, the model often selects one of two complementary literals as the defeasible fact. This behavior demonstrates that LLMs struggle with defeasible reasoning, which requires the management of inconsistency and incomplete information. The ability to handle such complexities is crucial for models engaged in tasks involving nuanced logical deductions where facts can be overridden or contradicted by more compelling evidence. Furthermore, another significant observation is that LLMs sometimes fail to identify all relevant pairs of arguments and counter-arguments. This limitation shows a basic problem with the models' ability to completely understand all the arguments in a DeLP problem.

6 Conclusion

In this paper, we investigate the formal defeasible reasoning capabilities of large language models using the DeLP framework. Our methodology involves creating a synthetic dataset with varying depths of reasoning to challenge the LLMs, and we introduced a Chain-of-Thought method to enhance their multi-step reasoning processes. Despite these efforts, our experiments demonstrate that LLMs struggle with managing defeasible information, highlighting a significant limitation in their ability to handle inconsistencies and incomplete knowledge effectively. This underscores a gap in the reasoning abilities of current LLMs compared to human-level intelligence, signaling the need for further research and development in this domain. The experimental analysis advances understandings of LLMs' capabilities in formal logical reasoning and sets the stage for further developments in this critical area of AI research.

References

[1] Besta, M., N. Blach, A. Kubicek, R. Gerstenberger, L. Gianinazzi, J. Gajda, T. Lehmann, M. Podstawski, H. Niewiadomski, P. Nyczyk and T. Hoefler, *Graph of Thoughts: Solving Elaborate Problems with Large Language Models*, AAAI (2024), pp. 17682–17690.

[2] Cao, R., Y. Wang, L. Gao and M. Yang, *Dictprompt: Comprehensive dictionary-integrated prompt tuning for pre-trained language model*, Knowledge-Based Systems **273** (2023), p. 110605.

[3] Chen, C., B. Liao and B. Wei, *Evidence-based argumentation and its incremental semantics*, in: *International Conference on AI Logic and Applications*, Springer, 2023, pp. 3–17.

[4] Chi, H. and B. Liao, *A quantitative argumentation-based automated explainable decision system for fake news detection on social media*, Knowledge-Based Systems **242** (2022), p. 108378.

[5] Clark, P., O. Tafjord and K. Richardson, *Transformers as soft reasoners over language*, in: *IJCAI*, 2020, pp. 3882–3890.

[6] Creswell, A., M. Shanahan and I. Higgins, *Selection-inference: Exploiting large language models for interpretable logical reasoning*, in: *ICLR*, 2023.

[7] García, A. J. and G. R. Simari, *Defeasible logic programming: An argumentative approach*, Theory and practice of logic programming **4** (2004), pp. 95–138.

[8] Han, S., H. Schoelkopf, Y. Zhao, Z. Qi, M. Riddell, L. Benson, L. Sun, E. Zubova, Y. Qiao, M. Burtell, D. Peng, J. Fan, Y. Liu, B. Wong, M. Sailor, A. Ni, L. Nan, J. Kasai, T. Yu, R. Zhang, S. Joty, A. R. Fabbri, W. Kryscinski, X. V. Lin, C. Xiong and D. Radev, *FOLIO: Natural language reasoning with first-order logic* (2022).

[9] Kazemi, M., Q. Yuan, D. Bhatia, N. Kim, X. Xu, V. Imbrasaite and D. Ramachandran, *BoardgameQA: A dataset for natural language reasoning with contradictory information* (2023).

[10] Kontopoulos, E., N. Bassiliades and G. Antoniou, *Visualizing semantic web proofs of defeasible logic in the dr-device system*, Knowledge-Based Systems **24** (2011), pp. 406–419.

[11] Koons, R., *Defeasible Reasoning*, in: *The Stanford Encyclopedia of Philosophy*, Metaphysics Research Lab, Stanford University, 2021, Fall 2021 edition .

[12] Pollock, J. L., *Defeasible reasoning with variable degrees of justification*, Artificial intelligence **133** (2001), pp. 233–282.

[13] Prakken, H., **32**, Springer Science & Business Media, 2013.

[14] Thimm, M., *The tweety library collection for logical aspects of artificial intelligence and knowledge representation*, Künstliche Intelligenz **31** (2017), pp. 93–97.

[15] Walton, D., "Methods of argumentation," Cambridge University Press, 2013.

[16] Wei, J., X. Wang, D. Schuurmans, M. Bosma, F. Xia, E. Chi, Q. V. Le, D. Zhou et al., *Chain-of-thought prompting elicits reasoning in large language models*, NeurIPS **35** (2022), pp. 24824–24837.

[17] Zhao, B., W. Jin, Y. Zhang, S. Huang and G. Yang, *Prompt learning for metonymy resolution: Enhancing performance with internal prior knowledge of pre-trained language models*, Knowledge-Based Systems **279** (2023), p. 110928.

www.ingramcontent.com/pod-product-compliance
Lightning Source LLC
Chambersburg PA
CBHW071137050326
40690CB00008B/1492